Ferrites - Synthesis and Applications

Edited by Maaz Khan

Published in London, United Kingdom

IntechOpen

Supporting open minds since 2005

Ferrites - Synthesis and Applications
http://dx.doi.org/10.5772/intechopen.95631
Edited by Maaz Khan

Contributors
Rashed Adnan Adnan Islam, Ravinder Dachepalli, Edapalli Sumalatha, Nyathani Maramu, Sadhana Katlakunta, Rajender Thota, Koteswari Gollapudi, Butreddy Reddy Ravinder Reddy, Shubha, Ankur Soam, Cesar Leandro Londoño Calderón, Oscar Moscoso Londoño, Felipe Ocampo Osorio, Jhon Augusto Jativa Herrera, Veerabhadrappa Jagadeesha Angadi, Krishtappa Manjunatha, Brian Jeevan Fernandes, Keralapura Parthasarathy Ramesh, muhammad Siyar, Asghari Maqsood, Sathyaseelan Balaraman, Baskaran Iruson, Senthilnathan Krishnmoorthy, Manikandan Elayaperumal

First published in London, United Kingdom, 2021 by IntechOpen
IntechOpen is the global imprint of INTECHOPEN LIMITED, registered in England and Wales, registration number: 11086078, 5 Princes Gate Court, London, SW7 2QJ, United Kingdom
Printed in Croatia

British Library Cataloguing-in-Publication Data
A catalogue record for this book is available from the British Library

Additional hard and PDF copies can be obtained from orders@intechopen.com

Ferrites - Synthesis and Applications
Edited by Maaz Khan
p. cm.
Print ISBN 978-1-83962-868-9
Online ISBN 978-1-83962-887-0
eBook (PDF) ISBN 978-1-83962-888-7

For EU product safety concerns:
IN TECH d.o.o., Prolaz Marije Krucifikse Kozulić 3, 51000 Rijeka, Croatia, info@intechopen.com or visit our website at intechopen.com.

Meet the editor

Dr. Maaz Khan is a senior researcher at PINSTECH, Pakistan. He obtained a Ph.D. and postdoctoral degree in Material Science. His research interests include the fabrication of nanomaterials and their structural, optical, magnetic, and electrical characterizations. He has authored more than 100 research articles and published nine books. Presently, he is editor in chief of *Journal of Materials, Processing and Design* and *The Nucleus,* and executive editor of *International Journal of Nano Studies & Technology*. He also serves as an editorial board member for several journals in the field of material science.

Contents

Preface

Ferrites are ceramic materials composed of iron(III) oxide mixed with small quantities of one or more metallic elements, such as cobalt, nickel, zinc, manganese, strontium, and barium. There are different types of ferrites, including spinel, garnet, ortho, and hexagonal ferrites. The crystallography, structural, electrical, and magnetic properties of ferrites depend upon the fabrication and chemical composition as well as the various heat treatments used during the preparation process of ferrites. Ferrites have unique electrical and magnetic characteristics that are useful for a wide range of technological applications including refrigerators, air conditioners, microwave ovens, radio and telecommunication devices, and computers.

The book includes seven chapters that focus on the fabrication and study of structural, magneto-electric, and dielectric properties of cobalt–erbium ferrites, nickel ferrite, manganese–zinc ferrite, zinc–copper ferrite, and graphene-based cobalt ferrite for microwave shielding. It also discusses applications of ferrites as electrode materials for supercapacitors. Moreover, the book examines nanoferrite-based drug delivery systems as adjuvant therapy for cancer treatment as well as the current and future challenges in this area. All the authors are experts and active in their respective research areas and thus this book provides the most current information. Proper references are provided at the end of each chapter that direct readers to the best sources in literature for more detailed information about the subject.

I am grateful to all the contributing authors for their help in completing this project. Many thanks to the staff at IntechOpen for making this book possible. I am particularly thankful to Author Service Manager Ms. Marica Novakovic for her affable and cooperative behavior throughout the publishing process. I hope this book helps readers obtain further knowledge about ferrites and their applications in different areas.

Maaz Khan
Pakistan Institute of Nuclear Science and Technology,
Pakistan

Section 1

Synthesis

Chapter 1

Synthesis and Study of Structural and Dielectric Properties of Dy-Ho Doped Mn-Zn Ferrite Nanoparticles

Krishtappa Manjunatha,
Veerabhadrappa Jagadeesha Angadi,
Brian Jeevan Fernandes and
Keralapura Parthasarathy Ramesh

Abstract

The Dy-Ho doped Mn-Zn Ferrite nanoparticles have been synthesized by solution combustion method using mixture of fuels as glucose and urea. The synthesized samples of structural properties were characterized through XRD (X-ray diffraction) and dielectric properties were studied through impedance analyzer. The XRD patterns of all samples confirms the spinel cubic structure having space group Fd3m. Further all synthesized samples reveal the single-phase formation without any secondary phase. The lattice parameters and hopping lengths were increases with increase of Dy-Ho concentration. SEM micrographs shows the porous nature for all samples. The crystallite size increases with increase of Dy-Ho concentration. The Dielectric properties of all the samples were explained by using Koop's phenomenological theory. The real part of dielectric constant, imaginary part of dielectric constant and dielectric loss tangent were decreases with increase of frequency. Th AC conductivity increases with increase of frequency. The real part of impedance spectra decreases with increase of frequency for all samples. The Cole-Cole plots shows the one semicircle for all samples. The high ac conductivity and low dielectric loss observed for all samples at high frequency region and this samples are reasonable for power transformer applications at high frequencies.

Keywords: Mn-Zn Ferrite nanoparticles, solution combustion method, Koop's phenomenological theory, Cole-Cole plots

1. Introduction

Nano-ferrites, which are currently being studied, have piqued curiosity on account of their remarkable electrical properties. Due to their extraordinary physical and chemical properties, spinel ferrites nanoparticles have become a significant field of research in nanotechnology, nanoscience, and nanoelectronics [1–6]. A kind of high resistance spinel ferrite with a conventional AFe_2O_4 formula, where A alludes to divalent (+2) metal ions. In deciding their significant applications, dielectric and electrical examinations of spinel ferrites assume a vital role. Doping has a considerable impact on the semiconductive property of spinel ferrites. The high electrical resistance of soft ferrites, which prevents undesirable eddy current losses in AC fields, is the most important asset they create for being qualified for high-frequency applications. Spinel ferrites might be utilized in a MCS (microwave communication system) [7], magnetic transmitter feeder [8], pulsed current monitor [9] and gas sensor [10]. Spinel ferrites, on the other hand, have excellent chemical stability and biocompatibility under physiological conditions [11]. Impedance spectroscopy was used to explore the electrical characteristics of spinel ferries. Electrical similar circuits with inductors, capacitors and resistors are commonly utilized models for complex impedance. A comprehensive impedance examination can provide the necessary information of a material's dielectric characteristics. This research enables for the separation of distinct total impedance contributions arising from bulk conductivity and interfacial phenomena, such as grain boundary, grain, and other electrode interface results.

Mn-Zn ferrites are relied upon to be mixed ferrites with Fe^{2+}/Fe^{3+} ions affecting dielectric characteristics at both A-site and B-site. As a result, Mn-Zn ferrites offer a wide range of electrical properties that can be applied to a wide range of technological applications, including telecommunications [12]. Few researchers are researching the effect of rare earth such as Sm, Gd, Eu, and Ce among others, on the varied properties of Mn-Zn ferrite, according to a thorough literature assessment [13, 14]. The dielectric properties of $Zn_{0.2}Ni_{0.8-x}Cu_xFe_2O_4$ (x = 0 to 0.6) can be enhanced by replacing Ni^{2+} with Cu^{2+}, according to Houshair et al. Rao et al. [15] examined on the cation distribution of Ni-Zn-Mn ferrite NPs. Bharamagoudar et al. [16] reported that the $Mn_{1-x}Zn_xFe_2O_4$ (where, x = 0, 0.25, 0.5, 0.75, 1) were prepared by solution combustion method and the dielectric constant decrements with enhancing of Zn content. In addition, Qian et al. [17] found that introducing Nd into Ni-Zn ferrite increased the dielectric properties. Impedance spectroscopy, in particular, has been carried out in various research. Rare earth (RE) metal ions (Dy&Ho) with larger ionic radii can cause crystal structure distortions [18]. As a result, replacing trivalent iron with RE metal ions at the Fe site improves dielectric and structural properties in Mn-Zn ferrites. There have been several studies on the integration of RE ions into Mn-Zn ferrites.

The main goal of this work is therefore to understand the dielectric constant, dielectric loss tangent, ac conductivity, cole-cole plot and impedance spectroscopy of Dy-Ho doped Mn–Zn ferrite. As indicated by the investigation accomplished, replacing of Fe^{3+} ions with a larger Dy^{3+}-Ho^{3+} ions results in a significant rise in dielectric and ac conductivity. In our current paper, we investigated the structure, dielectric properties of the current systems.

2. Synthesis method and characterizations

Stoichiometric quantity of metal nitrates such as manganese nitrate, zinc nitrate, ferrous nitrate, dysprosium nitrate, holmium nitrate and reducing agents as stoichiometry quantities of fuels glucose and urea were mixed in 30 ml distilled water, and the combined solution was taken in a borosil glass beaker. Then combined solution was continuously stirred for 60 min to achieve a homogeneous solution. At 450°C, this homogeneous solution was kept in a box style muffle furnace that had been preheated. The solution boils, froths, and then burns with a smoldering flame at first. The combustion process will be completed within 20 minutes. The flow chart of solution combustion method as shown in **Figure 1**.

The XRD was characterized by utilizing CuK_α radiation (λ = 1.5406 Å) and the 2θ diffractogram was run from 20° to 80° with a stage size of 0.02 We can deduce crystalline phase and structure from XRD patterns. The surface morphology of the all

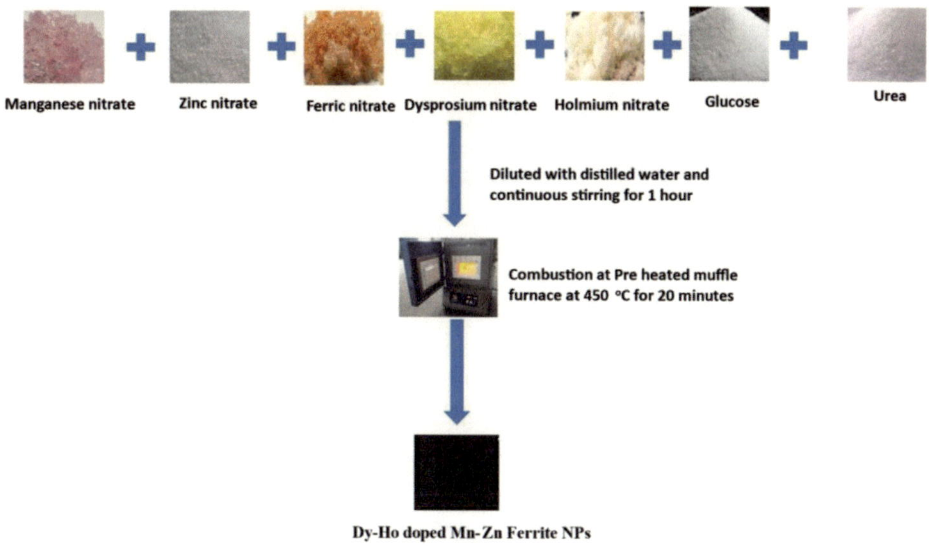

Figure 1.
Flow chart of solution combustion method for Dy-Ho doped M-Zn Ferrite NPs.

samples were analyzed by SEM images and the images were carried out by using JEOL (model JSM-840). For dielectric studies, the pellet of the sample was prepared using hydraulic press. The silver was pasted on it to get the electrical contact and heated in an oven for 2 hours at 55°C. The impedance spectroscopy measurement was performed in the frequency range up to 10 MHz using an Novocontrol Alfa A impedance analyzer.

3. Results

3.1 Structural analysis

The **Figure 2** depicts the XRD pattern of $Mn_{0.5}Zn_{0.5}Dy_xHo_yFe_{2-x-y}O_4$ (x = y = 0.005, 0.010, 0.015, 0.020, 0.025 and 0.030) NPs. The single-phase cubic structure was verified for all samples, and the pattern matched data card ICDD#10–0319 perfectly. The miller indices (hkl) suggested a spinel cubic structure without appearance of secondary phases. The lattice constant (a) values of were estimated by using the following relation [19].

$$a = \frac{\lambda\sqrt{h^2 + k^2 + l^2}}{2\,\sin\theta} \tag{1}$$

For x = y = 0.005 to 0.03 concentration, the values of 'a' were found 8.3964 to 8.4245 Å, respectively. Eq. (1) was utilized to estimate the crystallite size of $Mn_{0.5}Zn_{0.5}Dy_xHo_yFe_{2-x-y}O_4$ (x = y = 0.005, 0.010, 0.015, 0.020, 0.025 and 0.030) NPs using the Debye Scherrer Equation [20, 21];

$$D = \frac{k\,\lambda}{\beta\cos\theta} \tag{2}$$

The "λ" denotes the X-ray wavelength, the "β" denotes the FWHM value, k is the Scherrer constant and θ denotes the diffraction angle. The crystallite sizes measured were 11.88 to 6.44 nm for x = 0.005 to 0.03, respectively. Large ionic radius of rare-

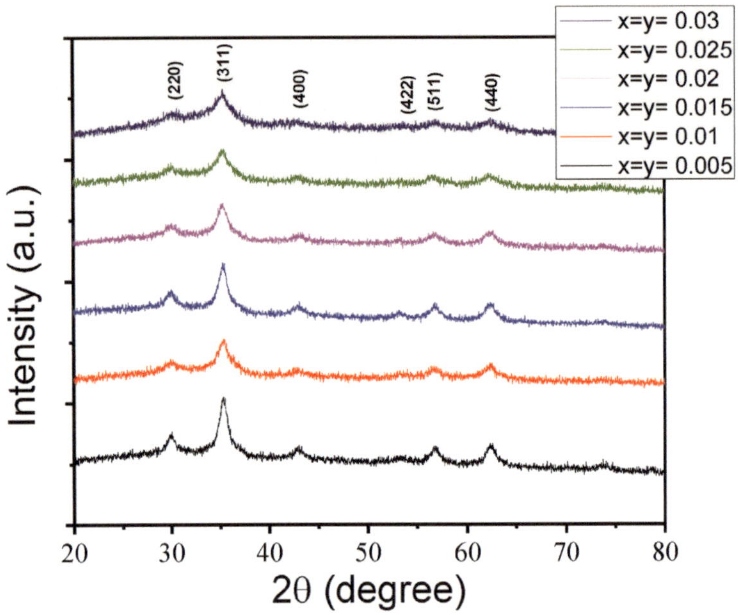

Figure 2.
The XRD patterns of $Mn_{0.5}Zn_{0.5}Dy_xHo_yFe_{2-x-y}O_4$ (x = y = 0.005, 0.010, 0.015, 0.020, 0.025 and 0.030) NPs.

earth ions increases the lattice parameter value while decreasing the average crystallite size, which is a popular trend [22]. However, in some cases, such as in our investigation, the researcher found different actions. The introduction of the Dy^{3+}-Ho^{3+} ions cause increases in the lattice parameter in our analysis. As the large ionic radius of Dy^{3+} (0.912 Å) and Ho^{3+} (0.901 Å) ions replaces the small ionic radius of Fe^{3+} (0.645 Å) ion at the B-site position, the lattice structure becomes asymmetric [23]. The hopping length at tetrahedral and octahedral sites was estimated by using following equations

$$L_A = \frac{\sqrt{3}a}{4} \text{ and } L_B == \frac{\sqrt{2}a}{4} \quad (3)$$

and observed the increase of hopping lengths with the increase of Ho^{3+} content as the lattice parameter increased gradually [24].

3.2 SEM analysis

SEM micrographs of $Mn_{0.5}Zn_{0.5}Dy_xHo_yFe_{2-x-y}O_4$ (x = y = 0.005, 0.010, 0.015, 0.020, 0.025 and 0.030) nanoparticles are shown in **Figure 3**. The existence of surface morphology with pores, holes, and on their surfaces can be seen in the figures. The development of the fuels during the combustion process resulted in the formation of the dry frothy powder. We are unable to measure grain size due to the porous nature of the samples. The micrographs show that the particles are agglomerated, showing that the magnetic nanoparticles in powder form have a strong connection [25].

3.3 Dielectric studies

3.3.1 Real part of dielectric constant

The variation of real part of dielectric constant (ε″) with applied frequency as shown in **Figure 4.** The ε' reduces as the frequency increases, stays constant at higher frequencies, and declines as the Dy^{3+} and Ho^{3+} content increases. This behavior could be explained by using Koop's theory. In the lower frequency zone, the electrons exchange between ions follows the applied electric field and is responsible for high value of ε' [26]. Due of high conducting grains, the ε' is

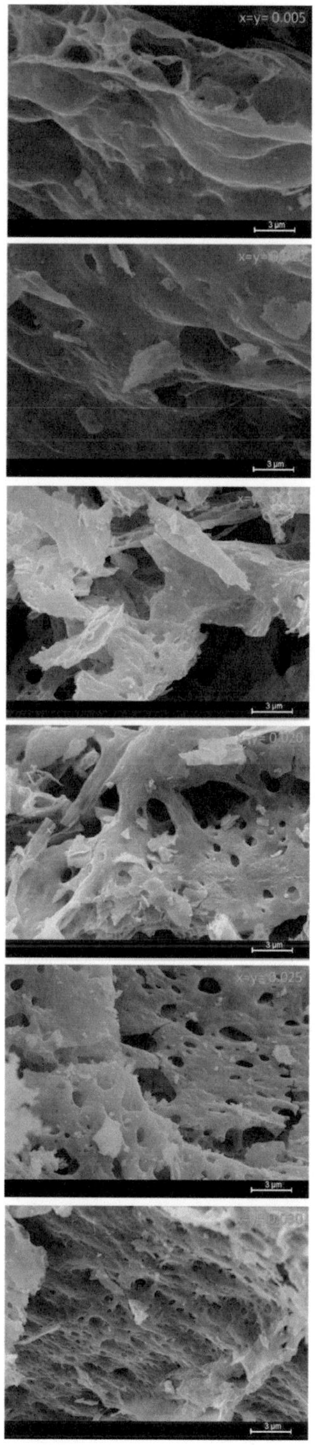

Figure 3.
SEM micrographs of $Mn_{0.5}Zn_{0.5}Dy_xHo_yFe_{2-x-y}O_4$ *(x = y = 0.005, 0.010, 0.015, 0.020, 0.025 and 0.030) nanoparticles.*

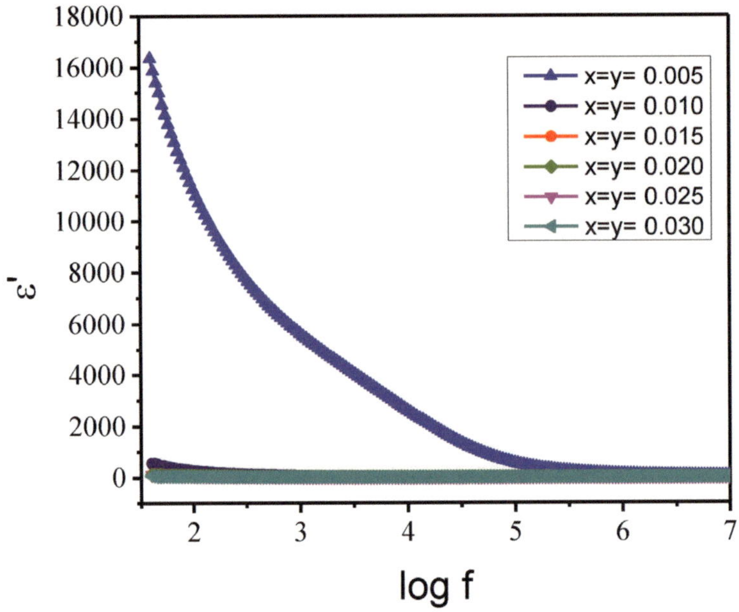

Figure 4.
The variation of real part of dielectric constant (ε′)with applied frequency of $Mn_{0.5}Zn_{0.5}Dy_xHo_yFe_{2-x-y}O_4$ *(x = y = 0.005, 0.010, 0.015, 0.020, 0.025 and 0.030) NPs.*

frequency independent at higher frequency region. The ionic and orientation polarizations weaken and eventually disappear as frequency rises, resulting in a drop in dielectric constant at higher frequency region [27]. Polarization is caused by electron exchange between Fe^{3+} and Fe^{2+} ions on the octahedral site in the ferrite lattice at lower frequencies.

3.3.2 Imaginary part of dielectric constant

The variation of real part of dielectric constant (ε″) with applied frequency as shown in **Figure 5**.

The concept of polarization and the hopping process can be used to understand the dielectric behavior of ferrite materials [28]. The following is the explanation for the observed dielectric loss in the ferrite samples: at lower frequency region the electron exchange between Fe^{2+} and Fe^{3+} is predominant and it follows the applied electric field. As the increase of frequency, the electron exchange between Fe^{2+} and Fe^{3+} ions does not follow the applied electric field.

3.3.3 Dielectric loss tangent

The variation of dielectric loss tangent (tan δ) with applied frequency as shown in **Figure 6**. Dielectric loss tangent in the ferrites is due to the lag of polarization with respect to the applied field [29, 30]. Ferrites with high tanδ are suitable candidates for the manufacturing of high frequency heating systems. Tanδ decreases with the applied frequency for each sample. This can be ascribed based on Koop's phenomenological model [31, 32]. At low frequencies region non conducting grain boundary gives maximum contribution for polarization. At lower frequency grain boundary contribution dominates results high resistivity and high value of dielectric loss tangent. Large quantity of energy is required for electron exchange

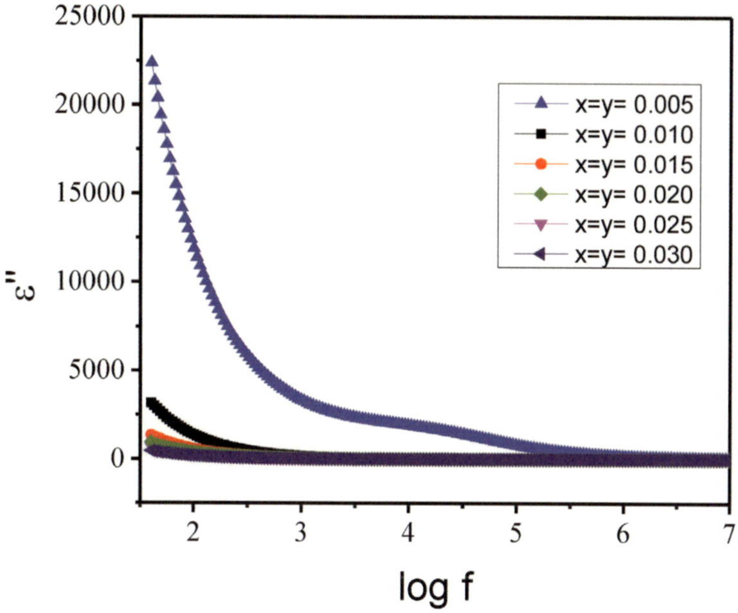

Figure 5.
The variation of imaginary part of dielectric constant with applied frequency of $Mn_{0.5}Zn_{0.5}Dy_xHo_yFe_{2-x-y}O_4$ (x = y = 0.005, 0.010, 0.015, 0.020, 0.025 and 0.030) NPs.

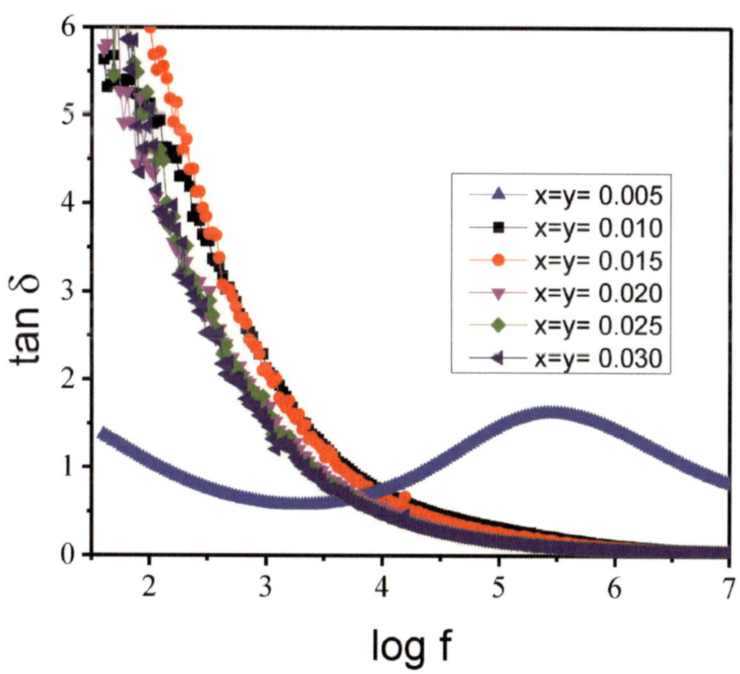

Figure 6.
The variation of dielectric loss tangent with applied frequency of $Mn_{0.5}Zn_{0.5}Dy_xHo_yFe_{2-x-y}O_4$ (x = y = 0.005, 0.010, 0.015, 0.020, 0.025 and 0.030) NPs.

between Fe^{3+} ions and Fe^{2+} ions at low frequency ensuing high value of loss tangent. At higher frequencies, small quantity of energy is enough for exchange of electron between Fe^{2+} and Fe^{3+} gives low resistivity and low value of loss tangent [33, 34]. At x = y = 0.005 concentration sample shows hump at mid of the frequencies, which was happened due to exchange of electron between ions frequency is matched with the applied frequency [35].

3.3.4 AC conductivity

The variation of AC conductivity (σ_{ac}) with applied frequency as shown in **Figure 7**. The frequency enhances with diminishing in σ_{ac} which can be explained due to hopping model. At lower frequency side independent of conductivity, so the σ_{ac} is small at lower frequency side. The Ho^{3+}-Dy^{3+} ions substitution on Fe^{3+} ions of B- site, here the electron exchange between ions and there is no electrons exchange between A site-B site. The electron exchange between A site-B site is most significant contrast with A site- A site and B site-B Site of spinel ferrite sample. The conduction mechanism enhances with enhancing the polarization there by enhancing the σ_{ac} [36].

3.3.5 Real part of impedance (Z′) and imaginary part of impedance (Z)

The variation of real part of impedance (Z′) with applied frequency as shown in **Figure 8**. The spectra unmistakably shows that the Z′ is diminishes with enhancing the frequency. Furthermore, because to the charge space polarization of the spinel ferrite sample [37], it remains constant at high frequency region. The imaginary part of impedance (Z″) varies with applied frequency, as shown in **Figure 9**. This spectrum (Z″ V/s log f) also named as loss spectrum. The frequency grows as Z″ decreases, and it reaches its maximum value at a certain frequency. The frequency then increases as Z″ decreases. Furthermore, the highest peak value rises as the concentrations of dysprosium and holmium rise. It results in the presence of relaxation time in the samples, which occurs as a result of space charge relaxation, which occurs when the sample is made up of grain borders and grain [38]. Furthermore, as the frequency shifts from low to high, the conduction mechanism shifts as well.

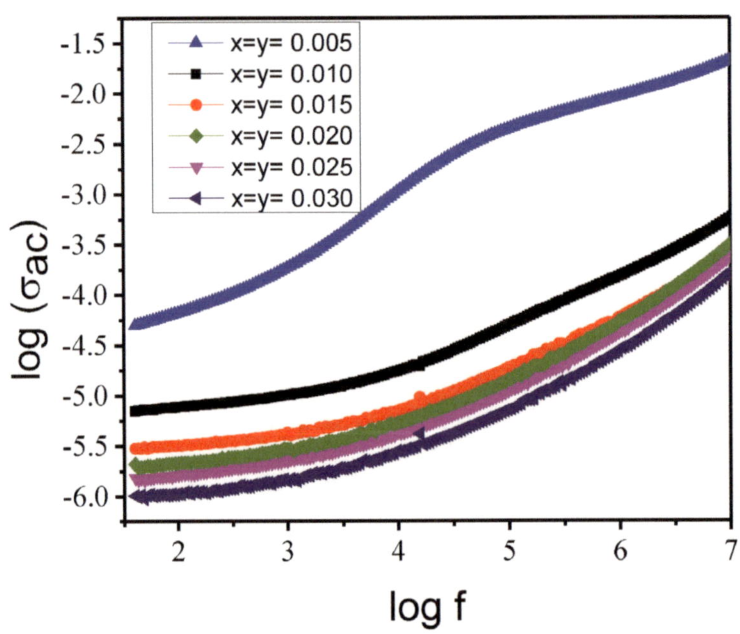

Figure 7.
The variation of AC conductivity with applied frequency of $Mn_{0.5}Zn_{0.5}Dy_xHo_yFe_{2-x-y}O_4$ (x = y = 0.005, 0.010, 0.015, 0.020, 0.025 and 0.030) NPs.

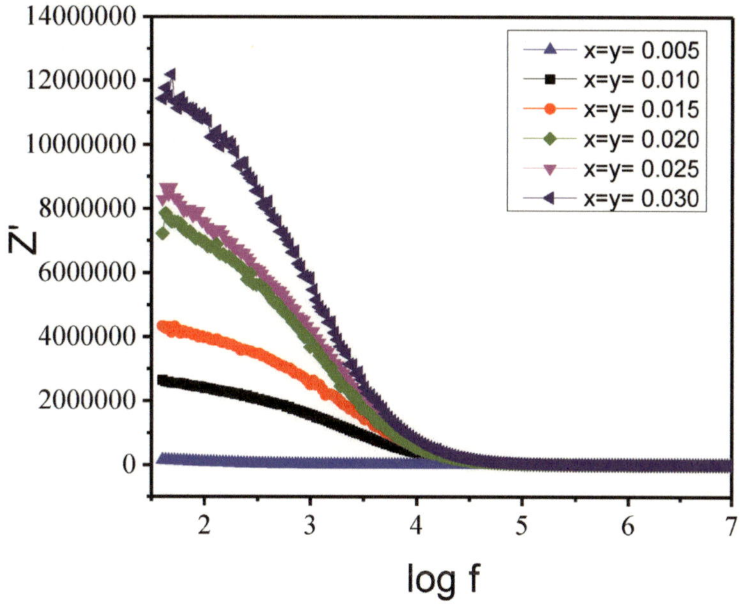

Figure 8.
The variation of real part of impedance (Z′) with applied frequency of $Mn_{0.5}Zn_{0.5}Dy_xHo_yFe_{2-x-y}O_4$ *(x = y = 0.005, 0.010, 0.015, 0.020, 0.025 and 0.030) NPs.*

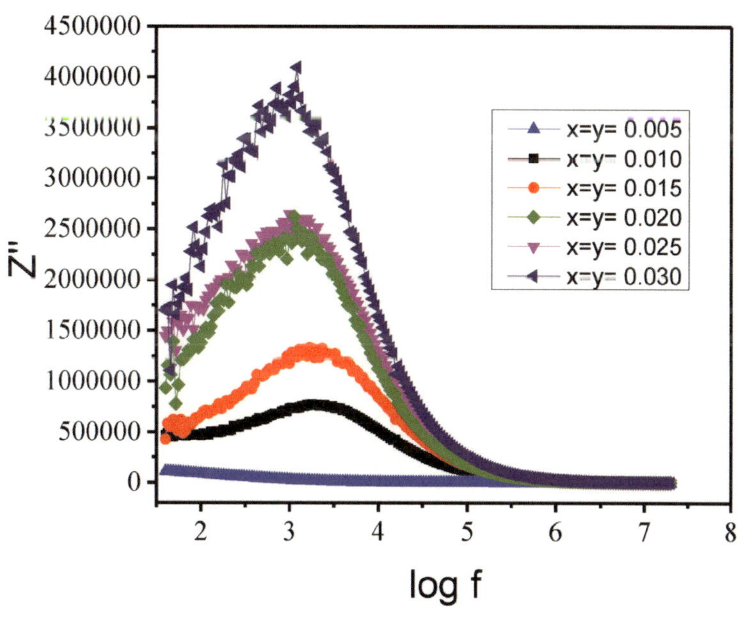

Figure 9.
The variation of imaginary part of impedance (Z) with applied frequency of $Mn_{0.5}Zn_{0.5}Dy_xHo_yFe_{2-x-y}O_4$ *(x = y = 0.005, 0.010, 0.015, 0.020, 0.025 and 0.030) NPs.*

3.3.6 Cole-Cole plot

The Cole-Cole plots (Z″ along y-axis and Z′ along y-axis) as shown in **Figure 10**. shows the and this plot is called Cole-Cole plots. The occurrence of a non-Debye kind of relaxation phenomenon in the Dy-Ho doped Mn-Zn ferrite NPs is confirmed by the Cole-Cole plots complex impedance spectra of the semicircle spectra.

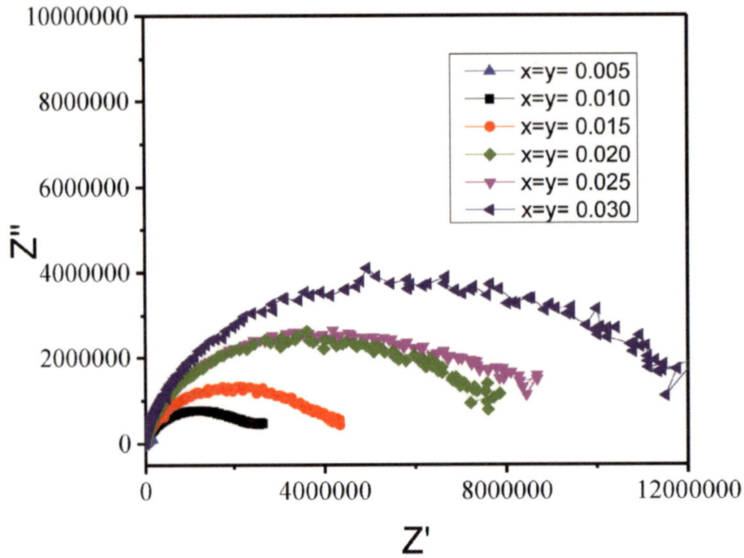

Figure 10.
Cole-Cole plots of $Mn_{0.5}Zn_{0.5}Dy_xHo_yFe_{2-x-y}O_4$ (x = y = 0.005, 0.010, 0.015, 0.020, 0.025 and 0.030) NPs.

Further, the maximum peak increases with increasing the Dy-Ho concentration. For the analogous circuit model, three series sets of capacitance and resistance are created in parallel. The complex impedance formula of an equivalent circuit is shown in Eq. (4) [39, 40].

$$Z = Z_0 + iZ'' = (1/R_b + i\omega C_b)^{-1} + (1/R_{gb} + i\omega C_{gb})^{-1} + (1/R_{el} + i\omega C_{el})^{-1} \quad (4)$$

Where R_b is the resistance of the material and C_b is the capacitance of the material, R_{el} and C_{el} is the contact impedance between material in the electrode. The capacitance and resistance assigned by C_{gb} and R_{gb}, respectively and brought about by the combination of grain boundary.

4. Conclusions

The synthesis of $Mn_{0.5}Zn_{0.5}Dy_xHo_yFe_{2-x-y}O_4$ (x = y = 0.005, 0.010, 0.015, 0.020, 0.025 and 0.030) NPs by solution combustion technique. The lattice parameters increases with increase of Dy-Ho content due to ionic radius of Dy^{3+} (0.912 Å) and Ho^{3+} (0.901 Å) ions greater than of Fe^{3+} (0.645 Å) ions. SEM micrographs shows the porous nature for all samples. The development of the fuels during the combustion process resulted in the formation of the dry frothy powder. The Dielectric properties of all the samples were explained by using Koop's phenomenological theory. The ε', $\varepsilon''$and tanδ were decreases with increase of frequency. Dielectric loss tangent in the ferrites is due to the lag of polarization with respect to the applied field. The AC conductivity rises as the frequency rises. For all samples, the real part of the impedance spectra diminishes as the frequency increases. Noticed that the maximum peak value increases with increase of dysprosium and holmium content in the imaginary part of impedance spectra. It gives a presence of relaxation time in the samples and it happened due to the space charge relaxation that overwhelms when the sample is composed of grain boundaries and grain. The appearance of a non-Debye type of relaxation phenomenon is linked to the presence of a single semicircle in the Cole-Cole plots for all samples. The high ac conductivity and low dielectric loss noticed for all samples at high frequency region are reasonable for power transformer applications at high frequencies.

Acknowledgements

Brian Jeevan Fernandes thanks University Grant Commission, Govt of India for the Dr. D.S. Kothari Post Doctoral Fellowship.

Author details

Krishtappa Manjunatha[1], Veerabhadrappa Jagadeesha Angadi[2]*, Brian Jeevan Fernandes[3] and Keralapura Parthasarathy Ramesh[3]

1 Department of Physics, School of Engineering, Presidency University, Bangalore, India

2 Department of Physics, P.C. Jabin Science College, Hubballi, India

3 Department of Physics, Indian Institution of Science (IISc), Bangalore, India

*Address all correspondence to: jagadeeshbub@gmail.com

References

[1] S. Amiri, H. Shokrollahi, The role of cobalt ferrite magnetic nanoparticles in medical science, Mater. Sci. Eng.: C, 33 (2013) 1–8.

[2] X. Meng, H. Li, J. Chen, L. Mei, K. Wang, X. Li, Mössbauer study of cobalt ferrite nanocrystals substituted with rare-earth Y^{3+} ions, J. Magn. Magn. Mater., 321 (2009) 1155–1158.

[3] J. Judith Vijaya, G. Sekaran, M. Bououdina, Effect of Cu^{2+} doping on structural, morphological, optical and magnetic properties of $MnFe_2O_4$ particles/ sheets/flakes-like nanostructures, Ceram. Int., 41 (2015) 15–26.

[4] J. Popplewell, L. Sakhnini, The dependence of the physical and magnetic properties of magnetic fluids on particle size, J. Magn. Magn. Mater., 149 (1995) 72–78.

[5] K. Manjunatha, I.C. Sathish, S.P. Kubrin, A.T. Kozakov, T.A. Lastovina, A. V. Nikolskii, K.M. Srinivasamurthy, Mehaboob Pasha, V. Jagadeesha Angadi, X-ray photoelectron spectroscopy and low temperature Mössbauer study of Ce^{3+} substituted $MnFe_2O_4$ J. Mater. Sci: Mater. Electron., 30 (2019) 10162-10171.

[6] K. Raj, B. Moskowitz, R. Casciari, Advances in ferrofluid technology, J. Magn. Magn. Mater., 149 (1995) 174–180.

[7] V.G. Harris, A. Geiler, Y. Chen, S.D. Yoon, M. Wu, A. Yang, Z. Chen, P. He, P. V. Parimi, X. Zuo, Recent advances in processing and applications of microwave ferrites, J. Magn. Magn. Mater., 321 (2009) 2035–2047.

[8] S.E. Jacobo, J.C. Aphesteguy, N.N. Shegoleva, G. V Kurlyandskaya, Structural and magnetic properties of nanoparticles of NiCuZn ferrite prepared by the selfcombustion method, in: Solid State Phenom., Trans Tech Publ,, 168 (2011) 333–340.

[9] R. Steiner, K. Merle, H.G. Andresen, A high-precision Ferrite-induction beam-current monitoring system, Nucl. Instruments Methods. 127 (1975) 11–15.

[10] A. Sutka, G. Mezinskis, A. Lusis, M. Stingaciu, Gas sensing properties of Zn-doped ptype nickel ferrite, Sensors Actuators B Chem., 171 (2012) 354–360.

[11] X. Wu, Z. Ding, N. Song, L. Li, W. Wang, Effect of the rare-earth substitution on the structural, magnetic and adsorption properties in cobalt ferrite nanoparticles, Ceram. Int., 42 (2016) 4246–4255.

[12] A. Verma, M.I. Alam, R. Chatterjee, T.C. Goel, R.G. Mendiratta, Development of a new soft ferrite core for power applications, J. Magn. Magn. Mater., 300 (2006) 500–505.

[13] V. Jagadeesha Angadi, K. Manjunatha, K. Praveena, Vinayak K. Pattar, Brian Jeevan Fernandes, S.O. Manjunatha, Jakeer Husain, S.V. Angadi, L.D. Horakeri, K.P. Ramesh, Magnetic properties of larger ionic radii samarium and gadalonium doped manganese zinc ferrite nanoparticles prepared by solution combustion method, J. Magn. Magn. Mater., 529 (2021) 167899.

[14] Salma Ikram, Jolly Jacob, K. Mahmood, A. Ali, N. Amin, U. Rehman, M. Imran Arshad, M. Ajaz un Nabi, Kashif Javid, A. Ashfaq, M. Sharif, S. Hussain, Influence of Ce^{3+} substitution on the structural, electrical and magnetic properties of $Zn_{0.5}Mn_{0.43}Cd_{0.07}Fe_2O_4$ spinel ferrites, Physica B Condens. Matter, 580 (2020) 411764.

[15] B.P. Rao, B. Dhanalakshmi, S. Ramesh, P.S.V.S. Rao, Cation distribution of Ni-Zn-Mn ferrite nanoparticles, J. Magn. Magn. Mater., 456 (2018) 444–450.

[16] R.C. Bharamagoudar, Jagadeesha Angadi, A.S. Patil, L.B. Kankanawadi, S. N. Mathad. Structural and Dielectric Properties of Combustion-Synthesized Mn-Zn Nanoferrites. Int. J Self-Propag. High-Temp. Synth. 28, 132–136 (2019).

[17] K. Qian, Z. Yao, H. Lin, J. Zhou, A. A. Haidry, T. Qi, W. Chen, X. Guo, The influence of Nd substitution in Ni–Zn ferrites for the improved microwave absorption properties, Ceram. Int., 46 (2020) 227–235.

[18] Asim Anwar, Sonia Zulfiqar, Muhammad Asif Yousuf, Sameh A. Ragab, Muhammad Azhar Khand, Imran Shakir, Muhammad Farooq Warsi, Impact of rare earth Dy^{+3} cations on the various parameters of nanocrystalline nickel spinel ferrite, J. Mater. Res. Technol., 9 (2020) 5313-5325.

[19] K. Manjunatha, V. Jagadeesha Angadi, R. Rajaramakrishna, U. Mahaboob Pasha, Role of 5 mol% Mg-Ni on the Structural and Magnetic Properties of Cobalt Chromates Crystallites Prepared by Solution Combustion Technique, J. Supercond. Nov. Magn., 33 (2020) 2861–2866.

[20] K. Manjunatha, K.M. Srininivasamurthy C.S. Naveen, Y.T. Ravikiran, E.I. Sitalo, S.P. Kubrin, Siddaling Matteppanavar, N. Sivasankara Reddy, V. Jagadeesha Angadi, Observation of enhanced humidity sensing performance and structure, dielectric, optical and DC conductivity studies of scandium doped cobalt chromate, J. Mater. Sci: Mater. Electron. 30 (2019) 17202-17217.

[21] K. Manjunatha, V. Jagadeesha Angadi, R.A.P. Ribeiro, M.C. Oliveira, S. R. de Lázaro, M.R.D. Bomio, S. Matteppanavar, S. Rayaprol, P. D. Babu, U. Mahaboob Pasha, Structural, Electronic and Magnetic properties of Sc^{3+} doped $CoCr_2O_4$ nanoparticles, New J. Chem., 44 (2020) 14246-14255.

[22] I.C. Sathisha, K. Manjunatha, V. Jagadeesha Angadi, Ranjeth Kumar Reddy, Structural, Microstructural, Electrical, and Magnetic Properties of $CuFe_{2-(x+y)}Eu_xSc_yO_4$ (where x and y vary from 0 to 0.03) Nanoparticles, J. Supercond. Nov. Magn., 33 (2020) 3963–3973.

[23] H.R. Lakshmiprasanna, K. Manjunatha, V. Jagadeesha Angadi, U. Mahaboob Pasha, Jakeer Husain, Effect of cerium on structural, microstructural, magnetic and humidity sensing properties of Mn–Bi ferrites, Nano-Struct. Nano-Objects, 24 (2020) 100608.

[24] I.C. Sathisha, K. Manjunatha, Anna Bajorek, B. Rajesh Babu, B. Chethan, T. Ranjeth Kumar Reddy, Y.T. Ravikiran, V. Jagadeesha Angadi, Enhanced Humidity Sensing and Magnetic Properties of Bismuth doped Copper ferrites for Humidity sensor Applications, J. Alloy Compd., 848 (2020) 156577 .

[25] M. Abhishek, K. Manjunatha, V. Jagadeesha Angadi, E. Melagiriyappa, B. N. Anandaram, H.S. Jayanna, M. Veena, K. Swaroop Acharya, Structural and magnetic properties of Eu3+ substituted Mg-Cd nanoferrites: A detailed study of Influence of high energy γ-rays irradiation, Chem. Data Coll., 28 (2020) 100460.

[26] K.M. Srinivasamurthy, K. Manjunatha, E.I. Sitalo, S.P. Kubrin, I.C. Sathish, S. Matteppanavar, B. Rudraswamy, V.J. Angadi, Effect of Ce^{3+} substitution on the structural, morphological, dielectric, and impedance spectroscopic studies of Co–Ni ferrites for automotive applications, Indian J. Phys., 94 (2020) 593–604.

[27] E. Melagiriyappa, H.S. Jayanna, B.K. Chougule, Dielectric behavior and ac electrical conductivity study of Sm^{3+} substituted Mg–Zn ferrites, Mater. Chem. Phys., 2008, 112, 68–73.

[28] R.C. Kambale, P.A. Shaikh, C.H. Bhosale, K.Y. Rajpure, Y.D. Kolekar, Dielectric properties and complex impedance spectroscopy studies of mixed Ni–Co ferrites, Smart Mater. Struct., 18 (2009) 085014.

[29] K. Manjunatha, V. Jagadeesha Angadi, M.C. Oliveira, S.R. de Lazaro, E. Longo, R.A.P. Ribeiro, S.O. Manjunatha, N.H. Ayachit, Towards shape-oriented Bi-doped CoCr2O4 nanoparticles from theoretical and experimental perspective: Structural, Morphological, Optical, Electrical and Magnetic properties, J. Mater. Chem. C, 9 (2021) 6452-6469.

[30] Somsack Vangchangyia, Ekaphan Swatsitang, Prasit Thongbai, Supree Pinitsoontorn, Teerapon Yamwong, Santi Maensiri, Vittaya Amornkitbamrung, Prinya Chindaprasirt, Very Low Loss Tangent and High Dielectric Permittivity in Pure-CaCu3Ti4O12 Ceramics Prepared by a Modified Sol-Gel Process, J. Am. Ceram. Soc., 95 (2012) 1497-1500.

[31] K. Manjunatha, V. Jagadeesha Angadi, K. M. Srinivasamurthy, Shidaling Matteppanavar, Vinayak K. Pattar and U. Mahaboob Pasha, Exploring the Structural, Dielectric and Magnetic Properties of 5 Mol% Bi^{3+}-Substituted $CoCr_2O_4$ Nanoparticles, J. Supercond. Nov. Magn., 33 (2020) 1747-1757.

[32] S.I.R. Costa, M. Li, J.R. Frade, D.C. Sinclair, Modulus spectroscopy of $CaCu_3Ti_4O_{12}$ ceramics: clues to the internal barrier layer capacitance mechanism, RSC Adv., 3 (2013) 7030-7036.

[33] K. Iwauchi, J. Appl. Phys., Dielectric properties of fine particles of Fe_3O_4 and some ferrites, 10 (1971) 1520–1528.

[34] V. Jagadeesha Angadi, H.R. Lakshmiprasanna, K. Manjunatha, Investigation of Structural, Microstructural, Dielectrical and Magnetic Properties of Bi^{3+} Doped Manganese Spinel Ferrite Nanoparticles for Photonic Applications, Bismuth - Fundamentals and Photonic Applications, IntechOpen (2020), ISBN: 978-1-83968-243-8. DOI: 10.5772/intechopen.92430.

[35] K.P. Padmasree, D.D. Kanchan, A.R. Kulkami, Impedance and modulus studies of the solid electrolyte system $20CdI_{2-80}$ $[xAg_{2O-y}$ $(0.7\ V_2O_5–0.3\ B_2O_3)]$, where $1\leq x/y\leq 3$, Solid State Ion. 177 (2006) 475.

[36] U. Ghazanfar, S.A. Siddiqi, G. Abbas, Study of room temperature dc resistivity in comparison with activation energy and drift mobility of NiZn ferrites, Mater. Sci. Eng. B 118 (2005) 132.

[37] T. Badapanda, S. Sarangi, S. Parida, B. Behera, B. Ojha and S. Anwar, Frequency and temperature dependence dielectric study of strontium modified Barium Zirconium Titanate ceramics obtained by mechanochemical synthesis, J. Mater. Sci. Mater. Electron., 26 (2015) 3069.

[38] K. Kamala Bharathi, J. Arout Chelvane and G. Markandeyulu, Magnetoelectric properties of Gd and Nd-doped nickel ferrite, J. Magn. Magn. Mater., 321 (2009) 3677-3680.

[39] K. Manjunatha, V. Jagadeesha Angadi, K.M. Srinivasamurthy, Shidaling Matteppanavar, Synthesis and Study of Structural, Dielectric Properties of $Co_{0.95}Bi_{0.05}Cr_2O_4$ nanoparticles, AIP Conf. Proc., 2274, 020004 (2020).

[40] Hafiz Muhammad, Tahir Farid, Ishtiaq Ahmad, Irshad Ali, Shahid M. Ramay, Asif Mahmood, G. Murtaza, Dielectric and impedance study of praseodymium substituted Mg-based spinel ferrites, J. Magn. Magn. Mater., 434 (2017) 143-150.

Chapter 2

Crystal Chemistry, Rietveld Analysis, Structural and Electrical Properties of Cobalt-Erbium Nano-Ferrites

Edapalli Sumalatha, Dachepalli Ravinder, Nyathani Maramu, Shubha, Butreddy Ravinder Reddy, Sadhana Katlakunta, Koteswari Gollapudi and Rajender Thota

Abstract

Synthesis of Cobalt-Erbium nano-ferrites with formulation $CoEr_xFe_{2-x}O_4$ (x = 0, 0.005, 0.010, 0.015, 0.020, 0.025, and 0.030) using technique of citrate-gel auto-combustion was done. Characterization of prepared powders was done by using XRD, EDAX, FESEM, AFM and FTIR Spectroscopy, DC resistivity properties respectively. XRD Rietveld Analysis, SEM, TEM and EDAX analysis were taken up in studying spectral, structural, magnetic and electrical properties. XRD pattern of CEF nano particles confirm single phase cubic spinal structure. The structural variables given by lattice constant (a), lattice volume (v), average crystallite size (D) and X-ray density(dx), Bulk density (d), porosity (p), percentage of pore space (P%), surface area (s), strain (ε), dislocation density (δ), along with ionic radii, bond length and hoping length were calculated. SEM and TEM results reveal homogeneous nature of particles accompanied by clusters having no impurity pickup. TEM analysis gives information about particle size of nanocrystalline ferrite while EDAX analysis confirm elemental composition. Emergence of two arch shaped frequency bands (ν_1 and ν_2) that represent vibrations at tetrahedral site (A) and octahedral site(B) was indicated by spectra of FTIR. The samples electrical resistivity (DC) was measured between 30°C -600°C with Two probe method. XRD Rietveld analysis confirm crystallite size lying between 20.84 nm–14.40 nm while SEM analysis indicate formation of agglomerates and TEM analysis indicate particle size ranging between 24 nm–16 nm. DC Electrical measurements indicate continuous decrease in resistivity with increasing temperature while increasing doping decreases curie temperature. The Magnetic parameters such as Saturation magnetization (M_s), Remanent magnetization (M_r), Coercivity (H_c) and Squareness ratio ($R = M_r/M_s$), Magnetic moment (n_B) were altered by doping of Er^{+3} content in the increasing order (x = 0.00 to 0.030). The increasing erbium content decreases magnetization thus converting the sample into soft magnetic material. Observations indicated strong dependence of magnetic properties on Erbium substitution and coercivity varies in accordance with anisotropy constant. Due to the presence of magnetic dipole Erbium substituted cobalt ferrites can be used in electromagnetic applications. The present study investigates the effect of different compositions of Er^{3+} replaced for Fe on structural properties and electrical resistivity of cobalt ferrites.

Keywords: Electrical Resistivity Properties, Co-Er nano-Particles, TEM, XRD, FESEM, EDAX, AFM

1. Introduction

Vigorous research has been accomplished on the fundamental, technological and potential applications of nano-ferrites. Nanomaterials of spinel ferrite have several applications in technology that include magnetic diagnostics and drug delivery [1], potential applications that include high density magnetic information storage devices [2], ferrofluid technology [3], magneto caloric refrigeration [4], magnetic recording media, magnetostriction [5], magnetic sensors, microwave devices and electrical generators etc. Ferrites are also used for catalyst and electronic devices. Ferrites are insulators exhibiting various magnetic and electric properties such as low electrical conductivity, dielectric loss, magnetic loss, relative loss factor, moderate dielectric constant, high initial permeability and saturation magnetization. Low eddy current and high resistivity makes ferrites better choice than metals [6]. Doping and thermal changes during synthesis and processing of cobalt-ferrites alter the distribution of metal ions influencing their structure and magnetic properties [7]. Priya et al. [8] doped Al ions, with Cobalt ferrite nano particles. They observed that Al doped cobalt ferrite to be suitable for high frequency applications and magnetic memory devices. Nasir Amin et al. [9] synthesized yttrium substituted cadmium ferrites. They reported yttrium doped Cd nanoferrites can be used in high-frequency microwave absorbing devices. Salma et al. [10] synthesized ferrite series having formulation SrYbyFe2–yO4 (y = 0.00 to 0.10). They observed the dispersion of frequency of ferrites is responsible for the natural magnetic resonance phenomenon and the domain wall pinning. As per the literature net magnetic moment of lanthanide series elements/ions depend on f-orbital electron number in which Er^{+3} is of small size (89 pm) with large magnetic moment (7 μB) [11]. Magnetic anisotropy of cobalt ferrites if doped get influenced by the existence of Er^{+3} because of strength in spin-orbit coupling. The present work reports the preparation and characterization of erbium doped cobalt ferrites combined by Citrate-gel auto combustion. The studies of $CoEr_xFe_{2-x}O_4$ with cobalt content x values ranging between 0.000 to 0.030 with step increase of x = 0.005 was reported. The crystallite size decrease with increasing erbium content indicating increase in surface area of the particle making it a good adsorbent. These adsorbents can be used in gas sensors and waste water treatment etc.

2. Experimental procedure

Synthesis of Cobalt-Erbium nano-ferrites with citrate-gel auto combustion technique was taken up with starting materials Cobalt Nitrate (Co $(NO_3)2{\cdot}6H_2O$), Ferricnitrate (Fe $(NO_3)3{\cdot}9H_2O$), Erbium Nitrate (Er $(NO_3).6H_2O$), Citric Acid ($C_6H_8O_7{\cdot}H_2O$) and Ammonia solution (NH_3) of 99.9% purity after weighing as per stoichiometric ratio. Later liquification of metal nitrates in distilled water was done and the mixture was stirred at 300 rpm for one hour to obtain a clear homogeneous solution. Next citric acid in aqueous form and metal nitrate was maintained in 1:3 ratio for all samples. Now, ammonia solution was added drop by drop to maintain Ph = 7. This solution on stirring was heated at 100°C temperature for ten to twelve hours to form a viscous gel. The water contained in the mixture gets evaporated slowly to form dry gel generating internal combustion to form a black colored desired sample. This sample was manually grinded and subjected to calcinations at 500°C in furnace for 4 hours. The step by step procedure for the synthesis of crystal ferrites is shown in the form of flow chart in **Figure 1**. Pellet was prepared with KBr hydraulic press (Model: M-15) in 2–3 mm thickness and 10 mm diameter size

Later these samples in pellet or powder form were used to characterize the material. Structural properties were analyzed with XRD (Bruker, CuKα, λ = 0.15406 nm), TEM (Model JEOL 2100F, Japan), Field-emission Scanning Electron Microscope (JEOL JSM-7600 F, Japan), Energy Dispersive X-ray Analyzer (EDAX) and, Atomic Force Microscopy (AFM: VEECO, USA). Two probe method was used to study electric properties.

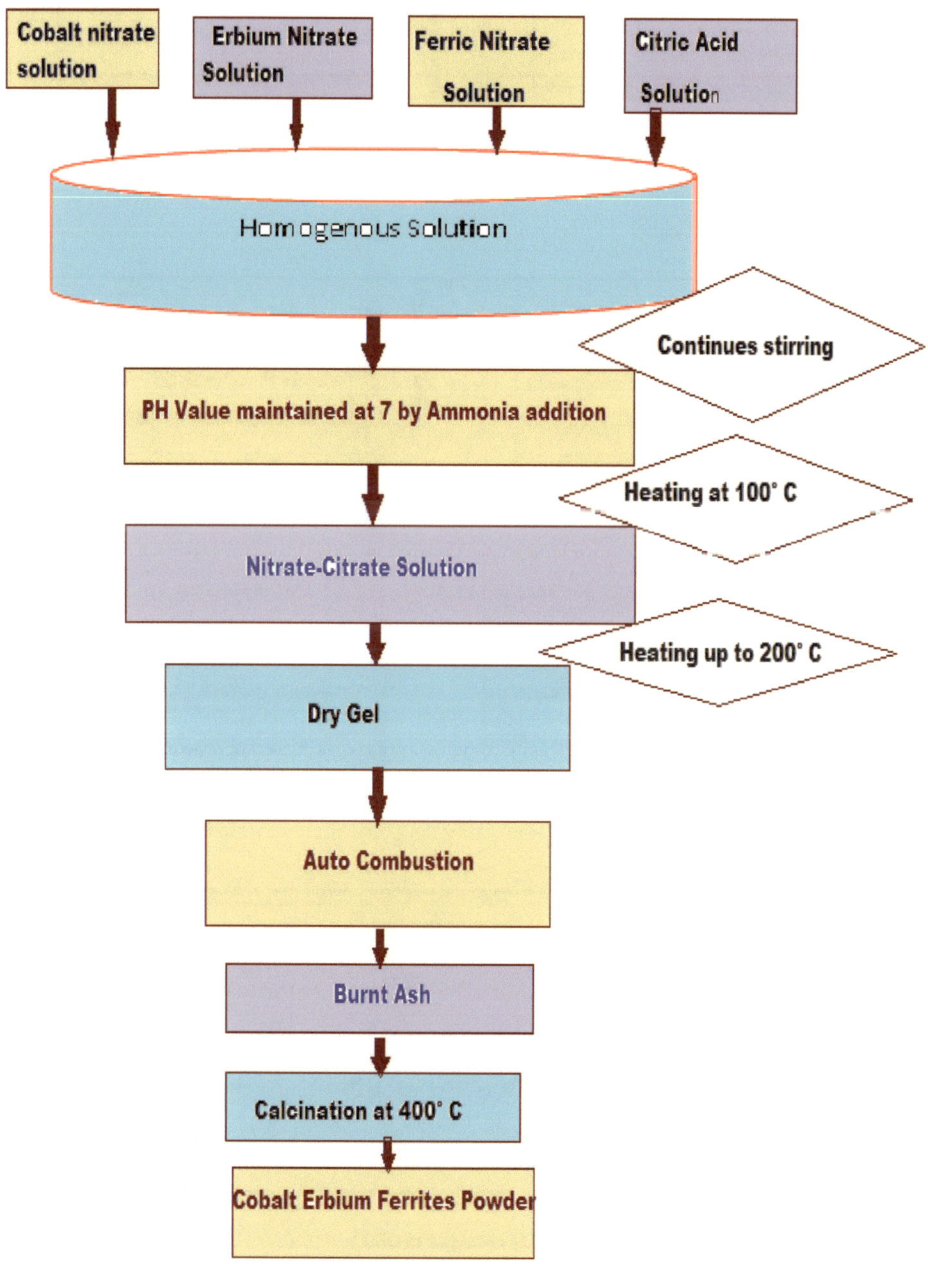

Figure 1.
Flow chart for the synthesis of cobalt-erbium ferrite using citrate-gel auto combustion technique.

3. Results and discussion

3.1 XRD analysis

Figure 2 displays the XRD Rietveld Refinement corresponding to samples of $CoEr_xFe_{2-x}O_4$ with values of x between 0.00 to 0.030 (x = incremented by 0.005).

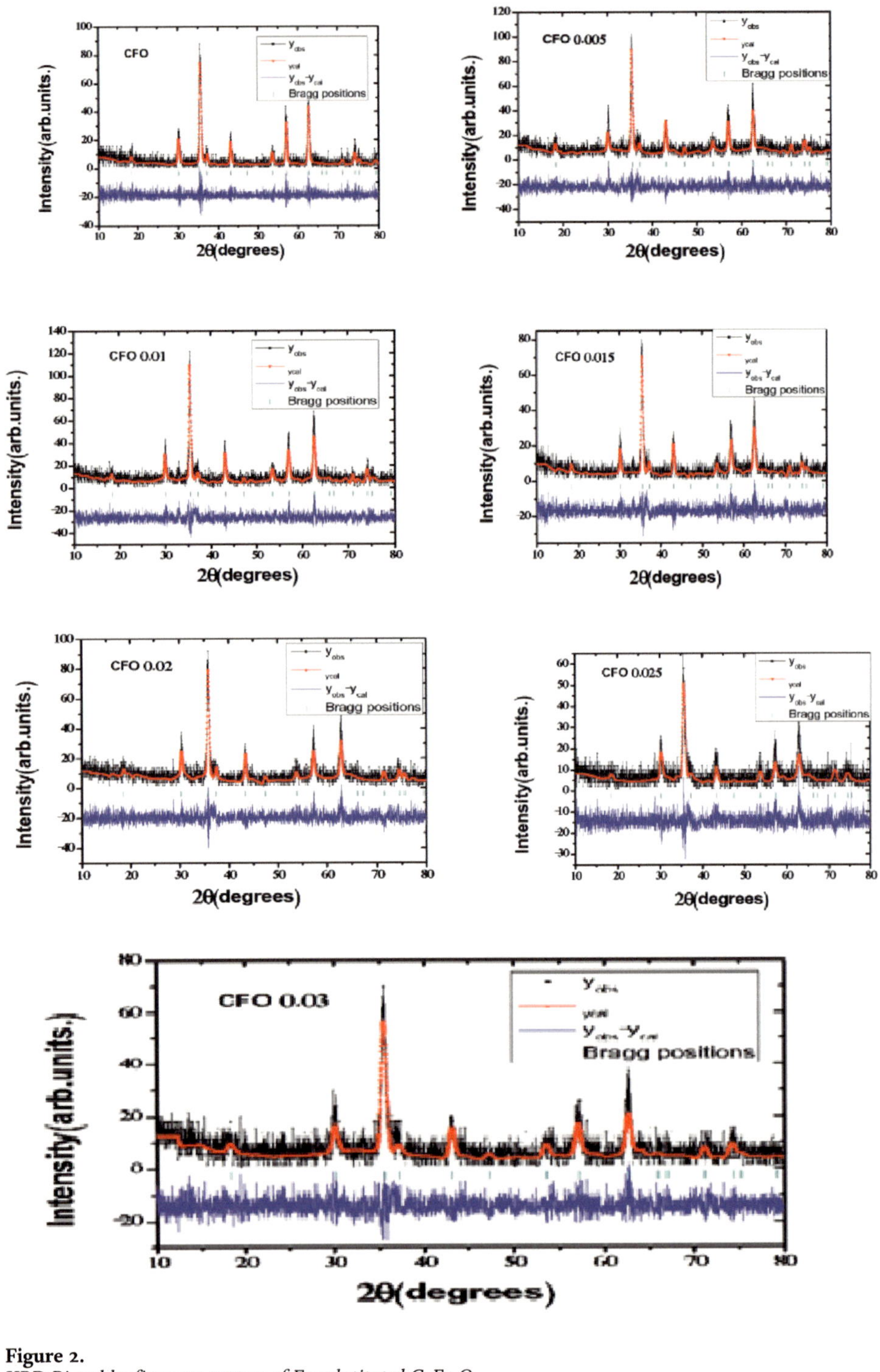

Figure 2.
XRD Rietveld refinement pattern of Er-substituted $CoFe_2O_4$.

It is observed that the peaks analogous to diffraction planes [111], [320], [311], [400], [511] and [440] match with usual data (JCPDS card no. 022–1086) confirming FCC cubic spinel structure for samples investigated [12, 13]. **Figure 3** shows shift in XRD peaks towards left hand side with increasing concentration of Er^{+3} ions in $CoFe_2O_4$ particles in concurrence with 'a' value. **Table 1** lists different parameters of XRD calculated for $CoEr_xFe_{2-x}O_4$ nanoparticles. The values of 'a' were calculated from the equation given [14].

$$a = d * \left(h^2 + k^2 + l^2\right)^{1/2} \tag{1}$$

where cell constant is given by 'a', inter planer spacing calculated from Bragg's equation ($2\,d\sin\theta = n\lambda$) is denoted by 'd' and miller indices are done by '*h,k,l*'.

It was reported that, low concentration RE (rare earth) doping in spinel ferrite experience phase separation and grain boundary diffusion giving rise to precipitation of additional crystalline phases like hematite (a-Fe_2O_3), metal monoxides and orthoferrites ($REFeO_3$) [15–17]. Hence in case of rare earth doped ferrites, Er^{+3}doped CFO having no impurity phase ($x \leq 0.010$) is exceptional and is because of auto-combustion. Induced effect due to substitution of erbium on the structure reflects two main observations given by decrease in size of crystal and increase in lattice constant both on small scale. The value of lattice constant slightly enhanced between 8.361 Å to 8.398 Å for x = 0.000 to x = 0.030 as per Law of Vegard [18]. Scherrer formula was used to calculate the crystallite size given by [19]:

$$L = \frac{0.9 * \lambda}{\beta Cos\theta} \tag{2}$$

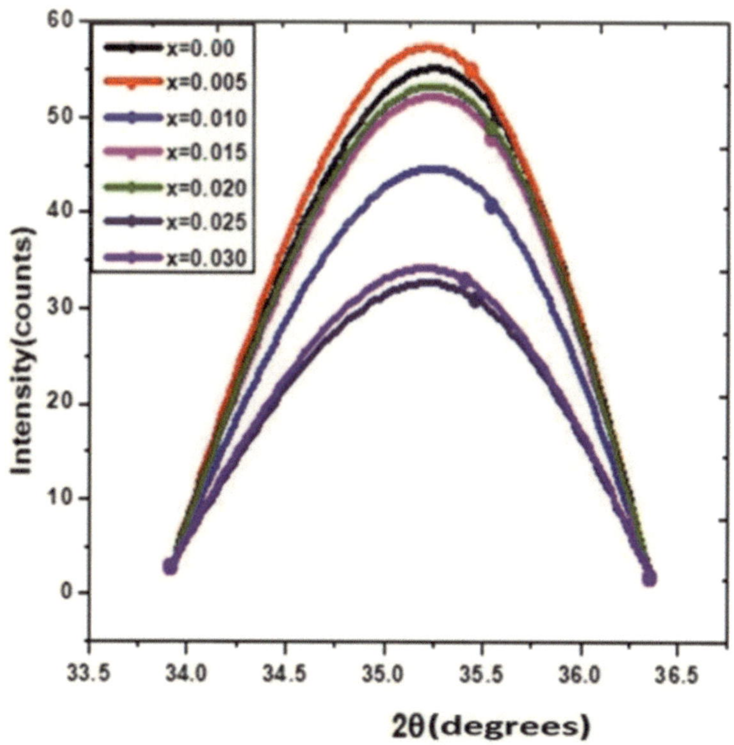

Figure 3.
XRD pattern of Er-substituted $CoFe_2O_4$ and shifting of peaks.

Compositions	Cell constant (Å)	Cell Volume V(Å^3)	Crystallite Size (nm)	X-ray density (dx) (gcm^{-3})	Bulk density(d) (gcm^{-3})	Porosity P (%)	Surface area(s) (m^2/gm)	Packing factor(P)	Strain (ε) x10^{-3}	Dis location density (δ)x10^{-4}
$CoFe_2O_4$	8.361	V = 584.48	20.84	5.3344	3.2113	39.8001	89.61	8.26	2.3	1.9
$CoEr_{0.005}$ $Fe_{1.995}$ O_4	8.367	V = 585.74	20.43	5.3356	3.2120	39.8005	91.39	8.09	2.3	2.0
$CoEr_{0.010}$ $Fe_{1.990}$ O_4	8.373	V = 587.00	19.19	5.3367	3.2127	39.7998	97.34	7.60	2.7	2.5
$CoE_{0.015}$ $Fe_{1.985}$ O_4	8.379	V = 588.26	19.02	5.3379	3.2134	39.8003	98.18	7.52	2.7	2.5
$CoEr_{0.020}$ $Fe_{1.980}O_4$	8.386	V = 589.74	17.73	5.3370	3.2129	39.7995	105.38	7.01	3.1	3.2
$CoEr_{0.025}$ $Fe_{1.975}O_4$	8.392	V = 591.01	15.56	5.3381	3.2135	39.8006	119.99	6.14	4.1	4.7
$CoEr_{0.030}$ $Fe_{1.970}$ O_4	8.398	V = 592.28	14.40	5.3392	3.2141	39.8018	132.58	5.68	4.8	5.9

Table 1.
Structural parameters of the prepared Co-Er nano ferrite sample.

where 'λ' = wavelength of x ray,'β' = peak width at half maximum height and constant 'K' = 0.9. The data related to intense peak (311) was used in estimating size (L). The results indicated reduction in size of crystallite from 20.84nmto14.40 nm (for x = 0.0 to 0.030). Further, the high intense peak (311) shifts towards the lower angle with increasing values of *x* (**Figure 3**). **Table 1** lists the physical parameters obtained from XRD which indicated increase in lattice constant of Co-Fe-Er spinel lattice which might be due to replacement of 8 small$Co^{2+}$and Fe^{3+} ions with big Er^{3+} ions. Huge difference in radii of these three ions induce strain during formation of lattice and diffusion processes. Requirement of more energy in absorbing RE^{3+} ions with more radii while replacing $Fe^{3+}$to form RE-O bond decreases crystallization energy and leads to particles of small size. Earlier literature reported similar results on RE-ion substituted cobalt ferrite [20–23]. From **Table 2**, EDAX confirmed the effect of incorporating Er^{3+} into CFO and stoichiometric amount of O, Fe, Co, and Er atoms. Therefore, XRD results are liable for expansion of unit cell due to larger Er^{3+} ion doping in CFO. Calculation of X-ray density (*Dx*) was done using [24]:

$$d_x = \frac{8 * M}{Na3} \tag{3}$$

Here
'*M*' = compositionmolecular weight.
'*N*' = Avogadro's number.
'*a*' = lattice constant.

X-ray density value is found to increase from 5.3344gm/cm3 to 5.3392gm/cm3 (x = 0.00 to x = 0.030) with increasing Er^{3+} content. The bulk density increased from 3.2113to 3.2141 (x = 0.00 to x = 0.030). At the same time, $CoFe_{2-x}Er_xO_4$ ceramics having more Er content (x = 0.015) exhibited lower $ErFeO_3$ orthoferrite amount along with primary spinel ferrite phase. Cobaltferrite in inverse spinel form has tetrahedral site occupied by half of Fe^{+3} while the remaining half of Fe^{+3} and Co^{-2} occupy octahedral sites [25]. Any change in site occupation of Fe^{+3} and Co^{-2} might be because of preparation technique and affect cell constant. Bulk densities were found from the relation [26].

$$d = \frac{m}{\pi r2h} \tag{4}$$

where pellets mass, thickness and radius are given by '*m*', '*h*' *and* '*r*'. Bulk densities exhibit inhomogeneous behavior due to pallets variable thickness and mass. The values of porosity in percentage were found using the relation.

Composition	$d_{A\text{-}A}$	$d_{B\text{-}B}$	$d_{A\text{-}B}$	$d_{A\text{-}OA}$	$d_{A\text{-}OB}$
$CoFe_2O_4$	3.620	2.956	3.466	1.926	2.025
$CoEr_{0.005}Fe_{1.995}O_4$	3.622	2.958	3.468	1.927	2.026
$CoEr_{0.010}Fe_{1.990}O_4$	3.625	2.960	3.471	1.928	2.027
$CoEr_{0.015}Fe_{1.985}O_4$	3.628	2.962	3.473	1.930	2.029
$CoEr_{0.020}Fe_{1.980}O_4$	3 631	2.964	3.476	1.931	2.031
$CoEr_{0.025}Fe_{1.975}O_4$	3.633	2.966	3.479	1.933	2.032
$CoEr_{0.030}Fe_{1.970}O_4$	3.636	2.969	3.481	1.934	2.033

Table 2.
Summarizes different bond lengths of A, B sites due to Er^{+3} ion doping in spinel lattice.

$$P\% = (1–d/d_x) \times 100 \quad (5)$$

Here d and d_x are apparent and experimental densities. Surface area was calculated by using the Eq. (16).

$$S = \frac{6000}{D * d} \quad (6)$$

Here, S = area of surface, D = crystallite size, d = Bulk density.
Strain was calculated by using the following Equation [27].

$$\text{Strain } (\varepsilon) = 1/D^\wedge 2 \quad (7)$$

Dislocation Density calculated by using following equation

$$\text{Dislocation density } (\delta) = 15\varepsilon/aD \quad (8)$$

Here ε is strain, a is lattice constant, D is crystallite size. Packing factor is calculated by using following equation

$$P = \frac{Lnm}{d} \quad (9)$$

Here L is crystallite size, d is inter planner spacing.

Cationic distributions that depend on factors like synthesis, total energy and thermal history are useful in understanding spinel ferrites behavior (electric and magnetic). Cationic calculations play important role in this regard. Average ionic radii of A, B sites were calculated from Stanley's equations:

$$r_A = [u - 1/4]a \times (3)^{0.5} - R_o \quad (10)$$

$$r_B = (5/8 - u)a - R_o \quad (11)$$

Here R_o is the radius of the oxygen ion (1.35 Å), 'u' is the oxygen parameter whose ideal value is 0.375Å and experimental value is 0.383Å.

Bonding lengths and hopping lengths are calculated by using following formulas [28].

Bonding lengths: Hoping lengths:

$$d_{A-A} = \frac{a}{4} \times (3)^{0.5} \quad (12)$$

$$d_{B-B} = \frac{a}{4} \times (2)^{0.5} a \quad (13)$$

$$d_{A-B} = \frac{a}{8} \times (11)^{0.5} \quad (14)$$

$$d_{A-OA} = a\left(u - \frac{1}{4}\right) * (3)^{0.5} \quad (15)$$

$$d_{B-OB} = a\left\{2\left(u - \frac{3}{8}\right)^2 + \left(\frac{5}{8} - u\right)^2\right\}^{0.5} \quad (16)$$

$$L_A = \frac{\sqrt{3}}{4} \times a \quad (17)$$

$$L_B = \frac{\sqrt{2}}{4} \times a \quad (18)$$

The difference in 'u' value in comparison with its ideal value on substituting Er^{+3} ions has been explained with r_A values. Increasing r_A values increase 'u' showing distortion in $CoFe_2O_4$ spinel lattice. Calculated values of ionic radii for B-sites are slightly higher than A-site because more Er^{+3} ions reside at B-site than A-site. Hopping length is the gap between magnetic ions at A, B sites. The hopping lengths between magnetic ions at A, B sites are denoted by L_A and L_B whose values reduce with addition of Er^{+3} content and is consistent with variation in lattice constant on adding Er^{+3} ions. The determined values from the formulas (10), (11), (17) and (18) are listed in **Table 1**.

By using the relations below structural parameters associated with A, B sites are calculated. Magnetic interactions and their strengths among AA, BB and AB sites mainly depend on bond length and bond angle existing between positive and negative ions. Increase in bond angle increases magnetic interaction strength while it reduces with increasing bond length as the strength has direct relation with bond angle and inverse relation with bond length. **Table 3** summarizes different bond lengths of A, B sites ($d_{A\text{-}A}$, $d_{B\text{-}B}$, $d_{A\text{-}B}$, $d_{A\text{-}OA}$, $d_{B\text{-}OB}$) which depict an increase in bond lengths of tetrahedral and octahedral sites which is due to Er^{+3} ion doping in spinel lattice which might be due to larger Er^{+3} ions replacing smaller Fe^{+3} ions.

3.2 EDAX analysis

Figure 4 displays the EDAX spectrums that analyzed elemental and atomic percentages of $CoFe_{2-x}Er_xO_4$ nanoparticles for x = 0.0, 0.005, 0.010, 0.015, 0.020, 0.025 and 0.030. It confirmed the presence of Co, O, Fe and Er. Er peak confirms Erbium substitution in the Fe_{2-x} lattice.

Table 2 summarizes atomic percentages of individual in $CoFe_{2-x}Er_xO_4$ nanoparticles. EDAX confirmed the effect of incorporating Er^{3+} into CFO and stoichiometric amount of O, Fe, Co, and Er atoms.

3.3 Field emission scanning Electron microscopy (FE-SEM)

Figure 5 shows studies on surface morphology of ferrite powders with the help of FE-SEM. The nature of ferrite particle in the samples is uniform indicating fine form of agglomeration and grain growth. Agglomerate formation specifies strong magnetic nature of erbium doped ferrites. These studies also confirm microstructure changes on doping Er^{+3}. A close look at these microstructures indicate improvement in microstructure and spherical shaped grains in all samples. Apart from this Erbium doping increases percentage of porosity in small range between 39.8001to 39.8018 illustrating individual grains and grain boundaries are separated.

Composition(x)	Co	Fe	Er	O
X = 0.000	12.40	34.43	0.000	26.86
X = 0.005	12.37	34.27	0.094	26.80
X = 0.010	12.34	34.10	0.18	26.74
X = 0.015	12.31	33.93	0.28	26.67
X = 0.020	12.28	33.77	0.37	26.61
X = 0.025	12.25	33.60	0.46	26.55
X = 0.030	12.23	33.44	0.55	26.49

Table 3.
summarizes atomic percentages of $CoFe_{2-x}Er_xO_4$ nanoparticles for x = 0.0, 0.005, 0.010, 0.015, 0.020, 0.025 and 0.030.

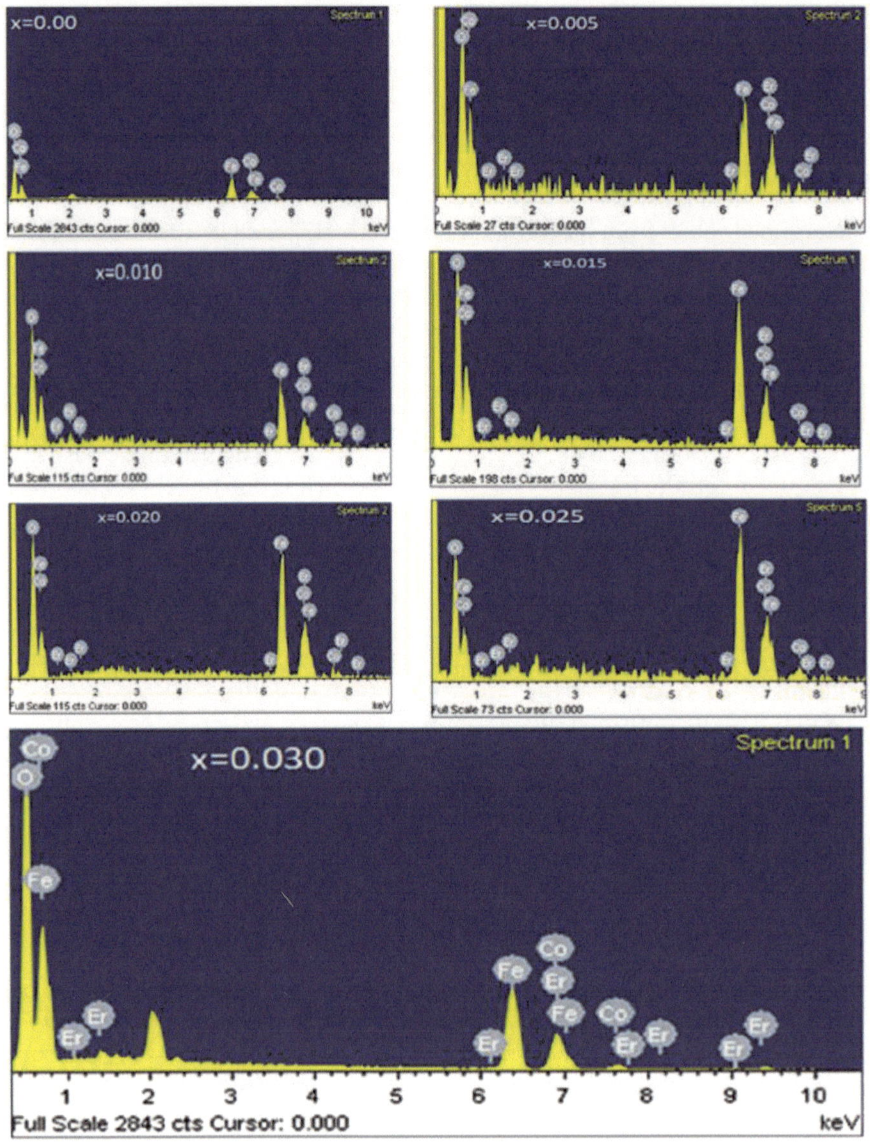

Figure 4.
Displays the EDAX spectrums that of $CoFe_{2-x}Er_xO_4$ nanoparticles.

3.4 Atomic force microscopy (AFM)

AFM was used to characterize the surface roughness of $CoEr_xFe_{2-x}O_4$ nano ferrite samples of the synthesized nanoparticles. The three-dimensional arrangement of the spherical nanoparticles and diameter are shown in **Figure 6**. The surface roughness increased when the coercivity increases, but in this work all the parameters crystallite size, saturation magnetization, remanent magnetization, coercivity decreased with the increasing of Er dopant from x = 0.00 to 0.030 in the cobalt ferrite. In view of the above, the largest surface roughness is observed for x = 0.0 sample and the lowest surface roughness is obtained for Er (x = 0.030) doped samples. This indicates that the surface activity of x = 0.0 ferrite has higher values compared to the range x = 0.005–0.030 ferrite samples. The largest surface roughness is observed for x = 0.0 sample that is, it behaves like hard ferrite and the

Figure 5.
Displays FE-SEM images of $CoFe_{2-x}Er_xO_4$ nanoparticles.

lowest surface roughness is obtained for Er (x = 0.030) doped samples. That is, it behaves like soft ferrite, hereby the ferrite is transformed from hard ferrite to soft ferrite due to the doping of Er content.

3.5 TEM analysis

Phase structure and morphology studies for the investigating synthesized samples were taken up through TEM analysis. **Figure 7** shows the TEM images and their

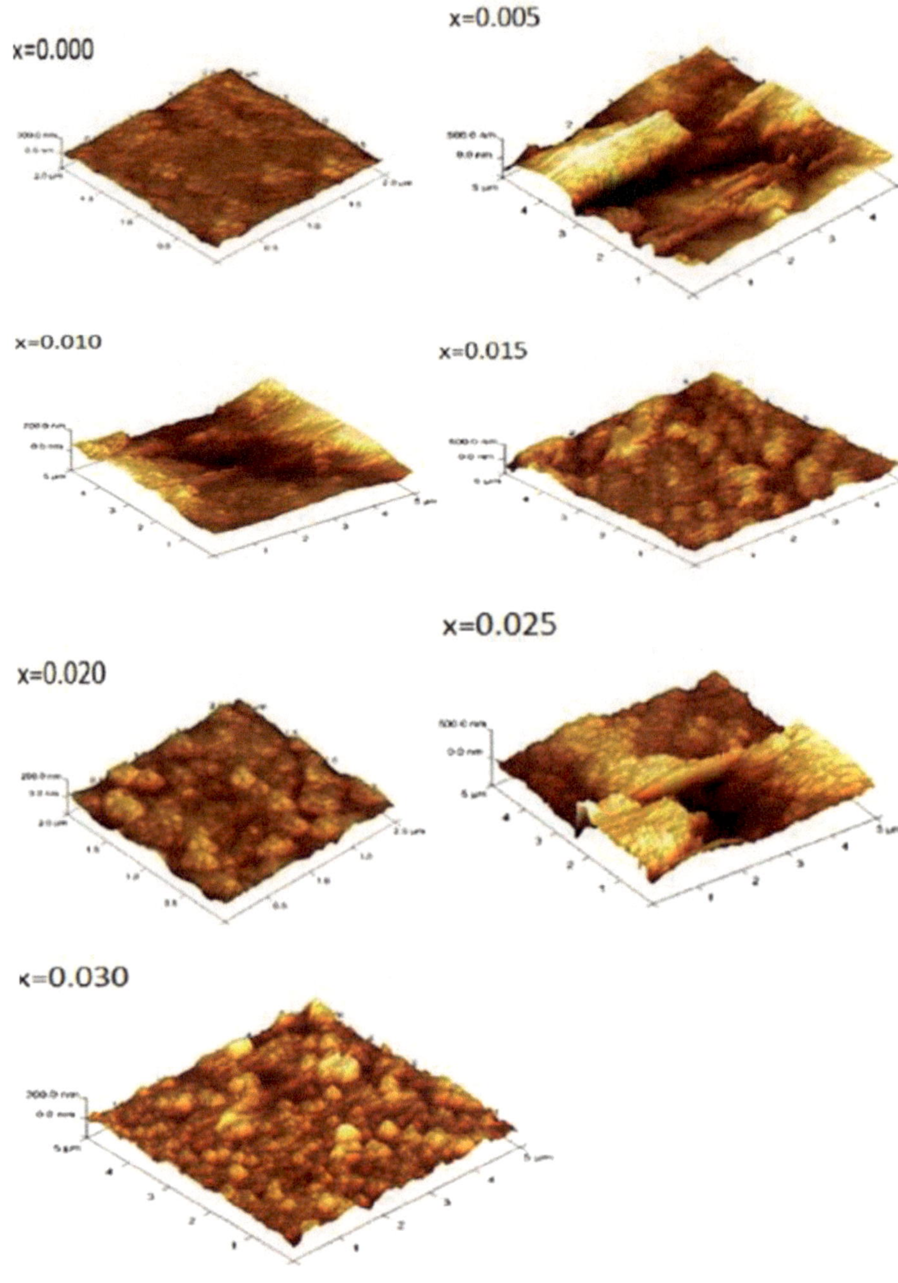

Figure 6.
AFM Micrographs of $CoFe_2O_4$ (x = 0.000) and $CoFe_{2-x}Er_xO_4$ (x = 0.005 to 0.030).

respective SAED images with particle size distribution chart of the samples got x = 0.0, 0.005, 0.01, 0.015, 0.02, 0.025 and 0.03 respectively. TEM and SAED images demonstrated spherical shape and less thickness for majority of the nanoparticles along with few elongated particles. Observation of TEM images confirm well distanced particles for lower concentration of Er^{+3} ions and increase in Er^{+3} ion substitution leads to agglomeration of particles because of magnetic nano particle interaction which makes the particles to be stacked on top of each other. The particle size measured from TEM images are in the range 16nm–24 nm.

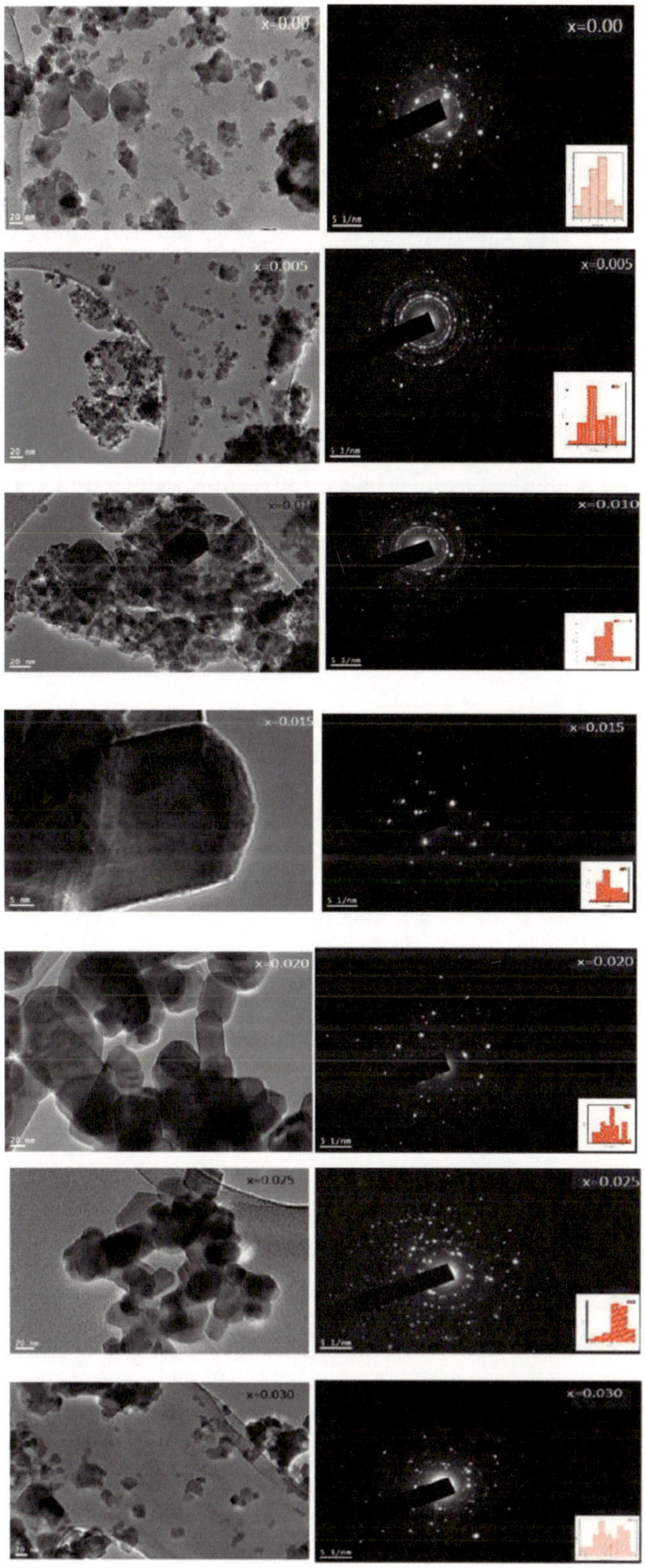

Figure 7.
TEM/SAED images of $CoFe_2O_4$ (x = 0.000) and $CoFe_{2-x}Er_xO_4$ (x = 0.005 to 0.030).

3.6 FTIR analysis

FTIR (Fourier Transform Infrared) spectroscopy is a very useful technique that estimates cationic redistribution at A and B sites of spinel ferrites. FTIR spectra for samples between 400 cm^{-1} and 1000 cm^{-1} was displayed by **Figure 8** in which two important broad bands (1 in the range 500 cm^{-1} – 600 cm^{-1} and 2 in the range 400cm^{-1} – 500 cm^{-1}) were observed. As per Waldron suggestion intrinsic vibrations of M–O complexes was shown by band 1 at site A site and band 2 at site B. This difference between 1 and 2 was because of variation in bond length of Fe^{+3}-O^{-2} at A, B sites [29]. Observations indicate shift in octahedral (2) and tetrahedral (1) bands towards higher frequency with the addition of Er^{+3} ions due to bond length variation, expansion in A, B sites and cation migration between two sites. The residency of Er^{+3} ions at B-site was also confirmed. FT-IR spectra of $CoFe_2O_4$ (x = 0.00) and $CoEr_xFe_{2-x}O_4$ (x = 0.005 to 0.030) nanoparticles are shown in **Figure 8**. The values of force constant at tetrahedral and octahedral (Ft&Fo) sites were determined using the formulas below [30] whose values are listed in **Table 4**.

$$Ft = 4\pi^2 c^2 \nu_1^2 u \tag{19}$$

$$Fo = 4\pi^2 c^2 \nu_2^2 u \tag{20}$$

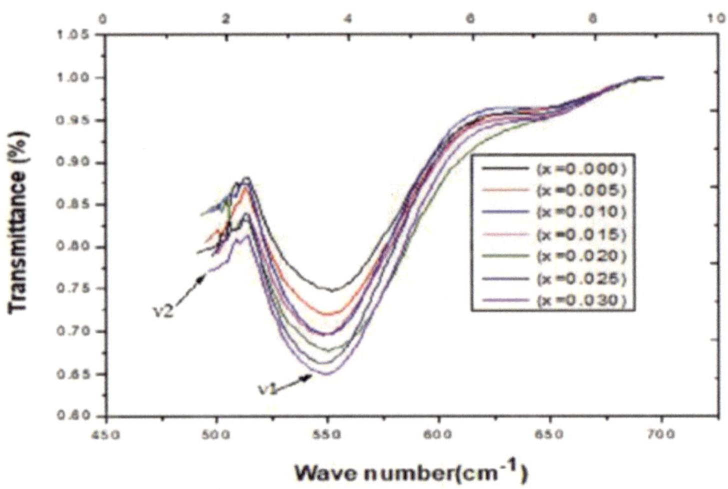

Figure 8.
FTIR spectra of $CoFe_2O_4$ (x = 0.000) and $CoFe_{2-x}Er_xO_4$ (x = 0.005 to 0.030).

Compositions	Wave number $v_1(cm^{-1})$	Wave number $v_2(cm^{-1})$	FT×105 (dynes/cm)	FO×105 (dynes/cm)
$CoFe_2O_4$	548	488	2.18	1.72
$CoEr_{0.005}Fe_{1.995}O_4$	550	490	2.20	1.74
$CoEr_{0.010}Fe_{1.990}O_4$	556	492	2.25	1.76
$CoEr_{0.015}Fe_{1.985}O_4$	564	494	2.31	1.77
$CoEr_{0.020}Fe_{1.980}O_4$	570	495	2.36	1.78
$CoEr_{0.025}Fe_{1.975}O_4$	572	496	2.38	1.79
$CoEr_{0.030}Fe_{1.970}O_4$	570	498	2.36	1.80

Table 4.
Summarizes FTIR modes(v_1,v_2) and force constants (F_T, F_O) of $CoFe_{2-x}Er_xO_4$ nano particles.

where vibrational frequencies of A, B sites are denoted by v_1, v_2, reduced mass of Fe^{3+} and O^{2-} ions is u, speed of light = c. Because of changes in bond lengths of Fe^{3+} and O^{2-} ions at A, B sites variation in values of force constant was determined.

3.7 Resistivity analysis

Resistivity figures signify distinct log ρ vs. 1000/T for various compositions of $CoFe_{2-x}Er_xO_4$. Resistivity reducing while increasing temperature, this behavior indicates that the semiconducting behavior of the prepared samples. Mobility of charge carriers (drift) reduces resistivity with temperature. Enhancement in temperature boosts enough energy to improve charge carriers hopping from one cationic site to other. The μD is growing up with the raise of Er^{+3} content the low drift mobility means temperature has not supplied sufficient potential to develop charge carriers to click from one site to another. Enrichment in μD with the boost of Er^{+3}contents advocate the enhancement of hopping from one cationic site to other for all nano ferrites synthesized particles. DC resistivity and drift mobility have inverse relation with each other. Observed resistivity figures indicated increase in resistivity initially for x = 0.000 and later decreases with increasing Er for x = 0.005 to 0.030. It is evident that all specimens contain a fixed quantity of Co. Resistivity vs. temperature curves of Er-substituted $CoFe_2O_4$ nanoparticles are shown in **Figure 9**. The resistivity is calculated from the following formula.

$$\rho = \frac{RA}{l} \tag{21}$$

Here R is the resistance, A is area of the pellet, l is length of the pellet.

3.8 Magnetic properties

M-H curves (Hysteresis Loops) are plots drawn between magnetization (M) and applied field (H) which helps us in analyzing magnetic response and magnetic parameters of ferrites under investigation. The M-H loops of all nanoparticles, that is $CoEr_xFe_{2-x}O_4$ (x = 0.00–0.030) heated at 500°C are displayed in **Figure 10**. The

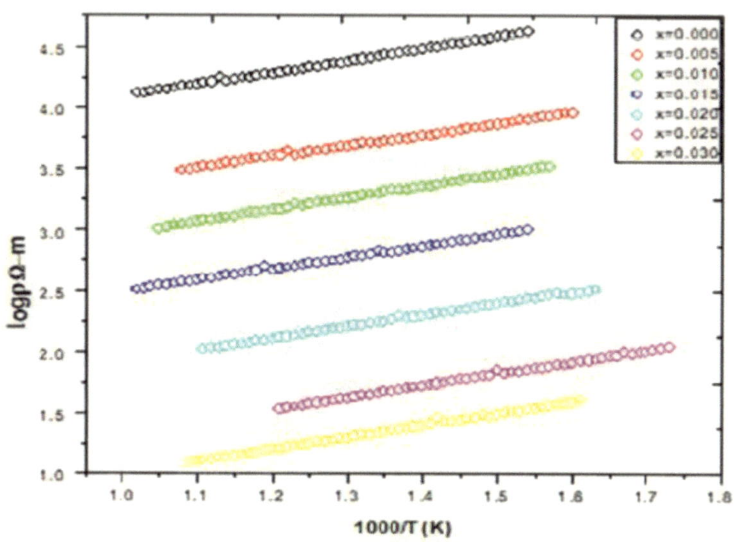

Figure 9.
Resistivity vs. temperature curves of Er-substituted $CoFe_2O_4$.

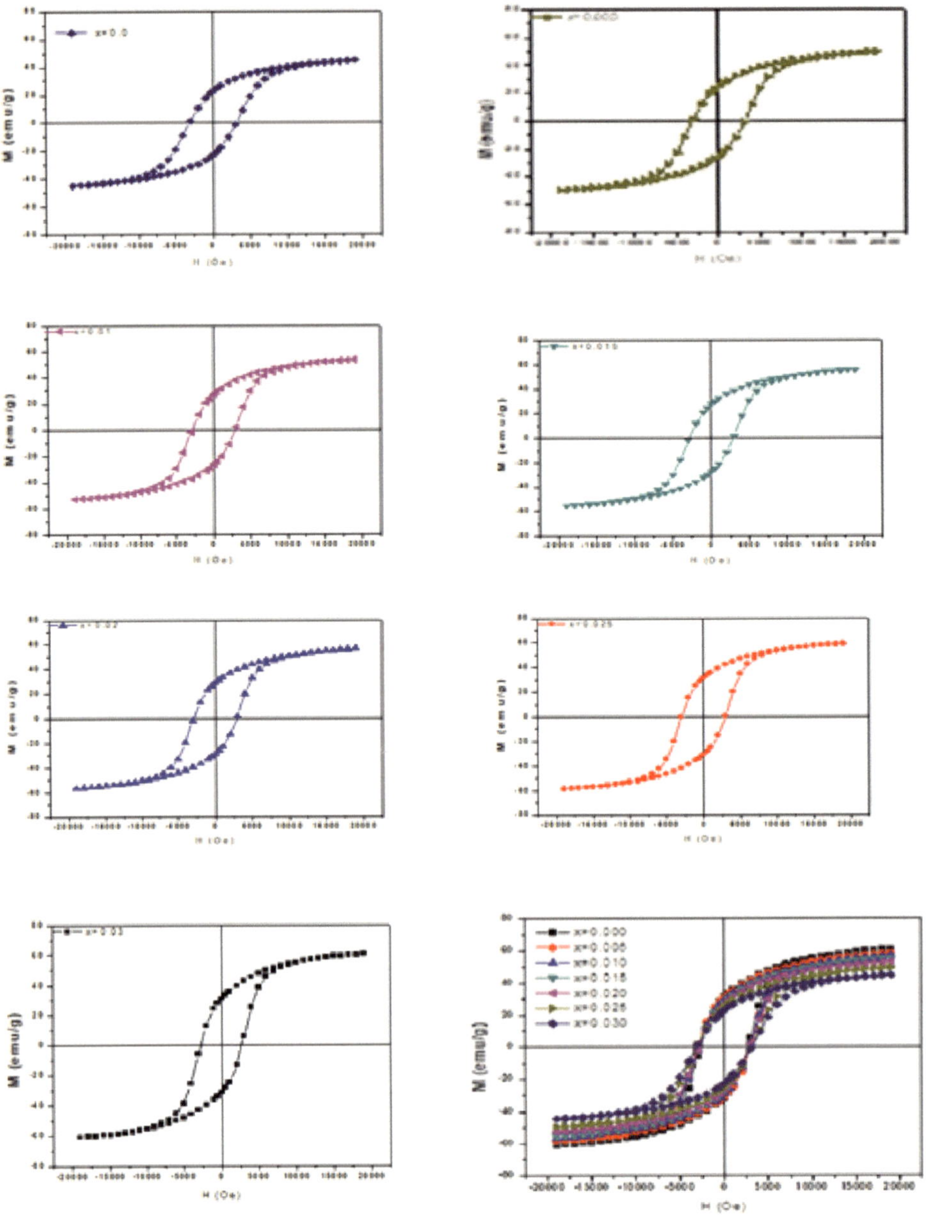

Figure 10.
The magnetic hysteresis curves of Er-substituted $CoFe_2O_4$ nano particles at room temperature.

measured magnetic parameters are displayed in **Table 5**. The Magnetic parameters such as Saturation magnetization (M_s), Remanent magnetization (M_r), Coercivity (H_c) and Squareness ratio ($R = M_r /M_s$), Magnetic moment (n_B) were altered by doping of Er^{+3} content in the increasing order (x = 0.00 to 0.030). Generally, dopant type, concentration and morphology will affect magnetic properties of soft ferrite sample. At the same time variation in magnetic parameters was seen due to microstructure with noting of higher saturation magnetization with higher grain size [31, 32]. **Table 5** indicate high saturation magnetization and coercivity due to large grain size in $CoFe_2O_4$ ferrites as depicted by the hysteresis loop in **Figure 10**. Ms. value decreased from 60 emu/g to 44 emu/g with decrease in grain size due to increased Er content in cobalt ferrite which may be due to increase of erbium

Composition	Lattice parameter (a)	Crystallite Size (nm)	Hc (c)	Ms (emu/g)	Mr (emu/g)	R=Mr/Ms	K (erg/Oe)	Magnetic moment (μB/f.u)
$CoFe_2O_4$	8.361	20.34	2998	60.6739	31.19	0.5141	189,479.783	2.5488
$CoEr_{0.005}Fe_{1.995}O_4$	8.392	20.43	2997	58.7486	31.07	0.5289	183,405.941	2.4738
$CoEr_{0.010}Fe_{1.990}O_4$	8.407	19.19	2996	56.9560	28.86	0.5067	177,750.433	2.4040
$CoEr_{0.015}Fe_{1.985}O_4$	8.367	19.02	2995	55.4902	27.66	0.5136	173,117.863	2.2784
$CoEr_{0.020}Fe_{1.980}O_4$	8.367	17.73	2993	53.1555	26.83	0.5208	165,723.033	2.1853
$CoEr_{0.025}Fe_{1.975}O_4$	8.386	15.56	2991	49.5845	25.15	0.5240	154.486.957	2.0400
$CoEr_{0.030}Fe_{1.970}O_4$	8.398	14.4	2989	44.8444	22.88	0.5275	139,625.157	1.8480

Table 5.
The Magnetic Properties of Er-substituted $CoFe_2O_4$.

cations in ferrite lattice site [33]. Particularly, high magnetic moment (5 μB) ferrite cations were replaced by erbium cations of magnetic moment 7 μB at B sites. In addition, increasing erbium cations may decrease ratio of ferric and ferrous ions at A, B sites thereby decreasing the magnetic exchange interaction between two sites [34] reducing the Msvalue. It was also observed that increase of erbium content reduced value of Hc from 18998 Oe to 18990Oeinitiating the fact that magnetic moment can be changed with low coercive field, hence coercivity variation is in agreement with variation in anisotropy constant. Henceforth, value of anisotropy constant 'K' will decrease further which decreases the energy of magnetic domain wall. Remanent magnetization values decreased from 31 emu/g to 22 emu/system supporting soft magnetic nature due to low coercivity in erbium doped cobalt ferrites [35]. **Table 5** indicates decrease in magnetic moment with increased erbium content which may be assigned to more probable chance of erbium cations to occupy B sites. As per the revealed data increasing erbium content decrease magnetization converting the sample into soft magnetic material. It is understood that increase in erbium content decreases value of 'K'. M–H loops indicated that soft magnetic Co-Er nano ferrites can be easily magnetized and demagnetized. Squareness ratio ($R = M_r/M_s$) was estimated from

$$R = \frac{M_r}{M_S} \tag{22}$$

where M_r is Remanent magnetization and M_s is saturation magnetization. Magnetic moment per unit (ηB) was calculated from [31, 32].

$$n_B = \frac{M_\omega \times M_S}{5585} \tag{23}$$

where Mω are samples molecular weight and saturation magnetization.

K (magnetic anisotropic constant) is related to the M_s (saturation magnetization) and Hc (magnetic coercivity) [28] by following relation

$$k = \frac{M_S \times H_c}{0.96} \tag{24}$$

4. Conclusions

Synthesis and characterization of erbium substituted cobalt ferrites along with conglomeration was done using citrate-gel auto combustion method. Significant induced effect of Erbium was observed on the structure of crystal structure, dielectric constant, morphology and electrical transport properties of cobalt ferrite material. Copy of secondary $ErFeO_3$ along with primary spinel cubic structure occur only for Er-content, x = 0.015,0.020 and regains its primary spinal structure for Er content x = 0.025,0.030 while the crystallite size decreased from 20.84 nm–14.40 nm. According to the SEM analysis the growth in grain along with agglomeration form was found for all samples. With the Erbium substitution which is a combined effect of decrease in resistivity. Small polaron hopping as well as thermally activated mobility of charge carriers was operative in CFEO ceramics and confirmed by DC electrical measurements. Observations indicated strong dependence of magnetic properties on Erbium substitution and coercivity varies in accordance with anisotropy constant. The presence of magnetic dipole could be useful for considering the Erbium substituted cobalt ferrites in electromagnetic applications. The studies of $CoEr_xFe_{2-x}O_4$ for compositions with cobalt content x = 0.0 to 0.030 with increasing order of x = 0.005 indicated decreasing crystallite size with increasing erbium content and increase in surface area of the particle makes it a good adsorbent. Hence these adsorbents can be used in gas sensors and waste water treatment etc. ...

Acknowledgements

Thanks to CSIR, New Delhi, India for Research Fellowship (CSIR-JRF). The authors are grateful to the Prof. Syed Rahman, Head Department of Physics, University College of Science, Osmania University Hyderabad for his constant encouragement.

Author details

Edapalli Sumalatha[1], Dachepalli Ravinder[1]*, Nyathani Maramu[2], Shubha[3], Butreddy Ravinder Reddy[2], Sadhana Katlakunta[2], Koteswari Gollapudi[1] and Rajender Thota[1]

1 Department of Physics, Osmania University, Hyderabad, Telangana, India

2 Department of Physics, University College of Science, Saifabad, Osmania University, Hyderabad, Telangana, India

3 Department of Electronics, St. Francis College for Women, Begumpet, Telangana, India

*Address all correspondence to: ravindergupta28@rediffmail.com

References

[1] Habiba AH, Ondeck CL, Chaudhary P, Bockstaller MR, McHenry ME . Evaluation of iron-cobalt/ferrite core-shell nanoparticles for cancer thermotherapy. Journal of Applied Physics. 2008; 103:07A307 https://doi.org/10.1063/1.2830975

[2] Song Q, Zhang ZJ. Shape Control and Associated Magnetic Properties of Spinel Cobalt Ferrite Nanocrystals. J. Am. Chem. Soc. 2004; 126, 19, 6164–6168 https://pubs.acs.org/doi/10.1021/ja049931r

[3] Vinosha PA, Das SJ. Investigation on the role of pH for the structural, optical and magnetic properties of cobalt ferrite nanoparticles and its effect on the photo-fenton activity. Materials Today: Proceedings 2018;5:8662-8671 https://www.sciencedirect.com/science/article/pii/S2214785317332807

[4] Vázquez-Vázquez C, Lovelle M, Mateo C, López-Quintela MA, Buján-Núñez MC, Serantes D, Baldomir D, Rivas J Magnetocaloric effect and size-dependent study of the magnetic properties of cobalt ferrite nanoparticles prepared by solvothermal synthesis. Wiley Online Library: 2008; 205: 1358-1362 https://doi.org/10.1002/pssa.200778128

[5] Nlebedim IC, Hadimani RL, Prozorov R, Jiles DC. Structural, magnetic, and magnetoelastic properties of magnesium substituted cobalt ferrite. Journal of Applied Physics 2013; 113: 17A928 https://doi.org/10.1063/1.4798822

[6] Prathapani S, Jayaraman TV, araprasadarao EK, Das D Structural and ambient/sub-ambient temperature magnetic properties of Er-substituted cobalt-ferrites synthesized by sol-gel assisted auto-combustion method, Journal of Applied Physics. Journal of Applied Physics, 2014;116:023908 https://doi.org/10.1063/1.4889929

[7] Cullity BD, Graham CD Introduction to Magnetic Materials, 2nd Edition. Wiley-IEEE Press, 2009;568 Pages https://www.wiley.com/en-in/Introduction+to+Magnetic+Materials%2C+2nd+Edition-p-9780471477419

[8] A. Sathiya Priya, D. Geetha, N. Kavitha, Effect of Al substitution on the structural, electric and impedance behavior of cobalt ferrite, Vacuum, 2019,Volume 160, Pages 453-460. https://doi.org/10.1016/j.vacuum.2018.12.004.

[9] Nasir Amin, Muhammad Sajjad Ul Hasan, Zahid Majeed, Zartashia Latif, Muhammad Ajaz un Nabi, Khalid Mahmood, Adnan Ali, Kiran Mehmood, Muneeba Fatima, Maria Akhtar, Muhammad Imran Arshad, Aisha Bibi, Muhammad Zahir Iqbal, Farukh Jabeen, N. Bano, Structural, electrical, optical and dielectric properties of yttrium substituted cadmium ferrites prepared by Co-Precipitation method, Ceramics International, 2020,Volume 46, Issue 13,Pages 20798-20809. https://doi.org/10.1016/j.ceramint.2020.05.079.

[10] Salma Aman, Naseeb Ahmad, M.H. Alhossainy, Hind Albalawi, Manal Morsi, Tahani I. Al-Muhimeed, Abeer A. AlObaid, Structural, Magnetic, Electrical and Microwave properties of Spinel Ferrites, Journal of Rare Earths, 2021, 1002-0721, https://doi.org/10.1016/j.jre.2021.04.015.

[11] Weiss A (1975) Craik (Ed.) DJ, Magnetic Oxides, Parts 1 and 2 (John Wiley & Sons, Bristol, 1975). Wiley Online Library, 1975;80:175-175 https://doi.org/10.1002/bbpc.19760800218

[12] Heiba ZK, Mohamed MB, Arda L, Dogan N Cation distribution correlated with magnetic properties of nanocrystalline gadolinium substituted nickel ferrite. Journal of Magnetism and Magnetic Materials, 2015; 391:195-202

https://doi.org/10.1016/j.jmmm.2015.05.003

[13] Kambale RC, Shaikh PA, Kamble SS, Kolekar YD, Effect of cobalt substitution on structural, magnetic and electric properties of nickel ferrite. Journal of Alloys and Compounds, 2009; 478:599-603 https://www.sciencedirect.com/science/article/abs/pii/S0925838808021257

[14] Dobrott RD Elements of X-ray Diffraction, Addison-Wesley, London. Characterization of Solid Surfaces, 1959;pp:147-178 https://link.springer.com/chapter/10.1007/978-1-4613-4490-2_8

[15] Andreu I, Natividad E, Ravagli C, Castroa M, Baldib G Heating ability of cobalt ferritenanoparticles showing dynamic and interaction effects. RSC Adv., 2014,4, 28968-28977 https://doi.org/10.1039/C4RA02586E

[16] Kakade SG, Ma Y, Devan RS, Kolekar YD, Ramana CV,Dielectric, Complex Impedance, and Electrical Transport Properties of Erbium (Er3+) Ion-Substituted Nanocrystalline Cobalt-Rich Ferrite. J. Phys. Chem. C 2016, 120, 10, 5682–5693 https://pubs.acs.org/doi/abs/10.1021/acs.jpcc.5b11188

[17] Park K, Hwang HK Fabrication and electrical properties of nanocrystalline Dy3+-doped CeO2 for intermediate-temperature solid oxide fuel cells. Energy, 2013; 55:304-309 https://doi.org/10.1016/j.energy.2013.04.017

[18] Kumar H, Singh JP, Srivastava RC, Negi P, Agrawal HM, Asokan K FTIR and electrical study of dysprosium doped cobalt ferrite nanoparticles, 2014. https://doi.org/10.1155/2014/862415

[19] Hashim M, Alimuddin, Kumar S, Shirsath SE, Kotnala RK, Shah J, Kumar R synthesis and characterization of nickel substuded cobalt ferrite. Materials Chemistry and Physics, 2012; 139:364-374 https://doi.org/10.1016/j.matchemphys.2012.09.019

[20] Wu Y, Li J, Bai H, He S, Hong Y, Shi K, Zhou Z Colossal Dielectric Behavior and Dielectric Relaxation of (Li, Fe) Co-Doped ZnO Ceramics. Wiley Online Library:Rapid Research Letter 2018; 12:1800126https://onlinelibrary.wiley.com/doi/abs/10.1002/pssr.201800126

[21] Salunkhe AB, Khot VM, Phadatare MR, Thorat ND, Joshi RS, Yadav HM, Pawar SH, Low temperature combustion synthesis and magnetostructural properties of CoMnnano-ferrites. Journal of Magnetism and Magnetic Materials, 2013; 352:91-98 https://doi.org/10.1016/j.jmmm.2013.09.020

[22] Rus SF, Vlazan P, Herklotz A Synthesis and characterization of Zirconium substituted cobalt ferrite nanopowders. Journal of Nanoscience and Nanotechnology 2016;16:1 https://www.osti.gov/pages/servlets/purl/1324165

[23] Rathore D, Kurchania R, Pandey RK Influence of particle size and temperature on the dielectric properties of CoFe2O4 nanoparticles. International Journal of Minerals, Metallurgy, and Materials, 2014; 21:408–414 https://link.springer.com/article/10.1007/s12613-014-0923-8

[24] Rahman MT, Ramana CV () Impedance spectroscopic characterization of gadolinium substituted cobalt ferrite ceramics. Journal of Applied Physics, 2014;116: 164108 https://doi.org/10.1063/1.4896945

[25] Goldman A Modern Ferrite Technology, 2nd ed. Ferrite Technology Pittsburgh, PA, USA 2006. https://download.e-bookshelf.de/download/0000/0009/50/L-G-0000000950-0002340313.pdf

[26] Raghasudha M, Ravinder D, Veerasomaiah P FTIR Studies and Dielectric Properties of Cr Substituted Cobalt Nano Ferrites Synthesized by Citrate-Gel Method. Nanoscience and Nanotechnology 2013, 3(5): 105-114 https://www.researchgate.net/publication/271907985_FTIR_Studies_and_Dielectric_Properties_of_Cr_Substituted_Cobalt_Nano_Ferrites_Synthesized_by_Citrate-Gel_Method

[27] Kurlyandskaya GV, Bhagat SM, Jacobo SE, esteguy JC, Schegoleva NN, Microwave resonant and zero-field absorption studies of pure and doped cobalt ferrite. Journal of Physics and Chemistry of Solids 2011;72:276-285 https://www.sciencedirect.com/science/article/abs/pii/S0022369711000126?via%3Dihub

[28] Kumari N, Kour S, Singh G, Sharma RK A brief review on synthesis, properties and applications of ferrites. AIP Conference Proceedings, 2020; 2220:020164 https://aip.scitation.org/doi/10.1063/5.0001323

[29] Pradeep A, Priyadharsini P, Chandrasekaran G, Sol–gel route of synthesis of nanoparticles of MgFe2O4 and XRD, FTIR and VSM study. Journal of Magnetism and Magnetic Materials 2008;320:2774-2779 https://www.sciencedirect.com/science/article/abs/pii/S0304885308007026

[30] Mazen SA, Abu-Elsaad NI IR Spectra, Elastic and Dielectric Properties of Li–Mn Ferrite. International Scholarly Research Notices, 2012:9 https://www.hindawi.com/journals/isrn/2012/907257/

[31] Kadam AA, Shinde SS, Yadav SP, Patil PS, Rajpure KY Structural, morphological, electrical and magnetic properties of Dy doped Ni-Co substitutional spinel ferrite. Journal of Magnetism and Magnetic Materials, 2013; 329:59-64 https://www.sciencedirect.com/science/article/abs/pii/S030488531200830X

[32] Ateia EE, Abdelmaksoud MK, Arman MM, Shafaay AS Comparative study on the physical properties of rare-earth-substituted nano-sized CoFe2O4. Applied Physics A, 2020; 126:91 https://scholar.cu.edu.eg/sites/default/files/drebtesam/files/comparative_.pdf

[33] Nag S, Roychowdhury A, Das D, Das S, Mukherjee S Structural and magnetic properties of erbium (Er^{3+}) doped nickel zinc ferrite prepared by sol-gel auto-combustion method. Journal of Magnetism and Magnetic Materials, 2018; 466:172-179 https://www.sciencedirect.com/science/article/abs/pii/S0304885317311472

[34] Pradeep A, Priyadharsini P, Chandrasekaran G Sol–gel route of synthesis of nanoparticles of MgFe2O4 and XRD, FTIR and VSM study. Journal of Magnetism and Magnetic Materials, 2008; 320:2774-2779 https://www.sciencedirect.com/science/article/abs/pii/S0304885308007026

[35] Gabal MA () Magnetic properties of NiCuZn ferrite nanoparticles synthesized using egg-white. Materials Research Bulletin, 2010; 45:589-593 https://www.sciencedirect.com/science/article/abs/pii/S0025540810000279

Chapter 3

Enhancement of the Magnetoelectric Effect in PZT-Ni Ferrite Composites Using Post Sintering Thermal Treatment

Rashed Adnan Islam

Abstract

In this chapter, the piezoelectric and magnetoelectric properties of Pb $(Zr_{0.52}T_{0.48})O_3$–X $NiFe_{1.9}Mn_{0.1}O_4$ (PZT-NFM, where X ranges from 3 to 10%) were investigated along with SEM and TEM analysis. Post sintering thermal treatment (annealing and aging) was introduced to enhance the piezoelectric and magnetoelectric property of the composites. The density was found around 7.6–7.8 gm/cc. The piezoelectric constant d_{33} and the voltage constant were found to be decreased with increasing magnetic particle from 3–10%. Experimental results indicated that the magnetoelectric voltage coefficient (dE/dH) increases with increasing percentage of Nickel Ferrite (~35 for 3% NF to 140 mV/Oe. cm for 10% NF at 1125°C). The samples, which were annealed and aged after sintering show as high as 50% (140 to 220 mV/cm Oe) increase in dE/dH coefficient. This is due to the increase in resistivity due to better homogeneity. The H-M curve of the composite powders show that the PZT-NFM is a ferromagnetic material as well. The SEM investigation shows very compact structure with average grain size around 150 nm and some areas of the NF rich phase. The X-ray mapping shows that the NF is distributed in the big grains of the matrix and after annealing it homogenizes.

Keywords: nickel ferrite, ferromagnetic, spinel, piezoelectric, magnetoelectric, perovskite

1. Introduction

Magnetoelectric materials contains combination of the ferroic properties such as polarization, and magnetization [1–4]. Due to the interrelationship between ferroelectricity and magnetism, it is possible to control of ferroelectric properties using magnetic field and vice versa. Cr2O3, BiFeO3, YMnO3 etc. which are single phase magnetoelectric (ME) materials do not provide much benefits due to very weak magnetoelectric effect [5–7]. The solution to this issue is composites utilizing the product property of the two materials and have much better ME voltage coefficient [8–10]. Magnetoelectric (ME) particulate sintered composite can be fabricated in combination of magnetostrictive and piezoelectric phases [11–15]. Among the many advantages of sintered composites, simplicity in synthesis, cost-effective materials and fabrication process, and better control of desired geometry are the most important ones. However, the ME effect in particulate sintered composite is still low and it is of the order of 100 mV/cm.Oe. Laminated composites which are fabricated by bonding colossus piezoelectric and magnetostrictive materials are very popular because they display excellent ME properties [16–19]. Dong et al. have shown a large response of 22 V/cm.Oe at 1 Hz using piezo fiber laminated between the high-permeability magnetostrictive FeBSiC alloy with epoxy [20]. The inherent materials property is the ferroelasticity that complemented both the ferroelectricity and ferromagnetism in same materials. This is a three-step mechanism: (a) ferroelectric ion movement needs structural building blocks, (b) super exchange type magnetic-interaction conduits and (c) the symmetry condition is satisfied [21]. One simple inference from this is that it is possible to synthesize the ferroelectric ferromagnets by replacing diamagnetic ions by paramagnetic ones on the B-site of oxyoctahedral ferroelectric perovskites.

Among the piezoelectric materials, Lead Zirconate Titanate (PZT) is the most popular for its high piezoelectric property and most importantly it shows magnetoelectric effect when used in a composite as reported by others. PZT when mixed with some magnetostrictive particles (such as ferrites) have been found to exhibit an extrinsic magnetoelectric effect resulting from a coupling interaction. The precondition to it is increased resistivity, almost zero to no interdiffusion, no chemical reaction and reduced interfacial such as microcracks and porosities [22]. Both the magnitude of magnetostriction and the slope of the magnetostriction curve with respect to applied magnetic field should have a large value in order to achieve high values of pseudo-piezomagnetic coefficients in the magnetostrictive phase [22–24]. In this regard, Ni ferrite doped with Mn ($NiFe_{1.9}Mn_{0.1}O_4$) can be promising candidate due to its increased resistivity, superior magnetization and small coercive field in order to switch the domains.

Co-firing of piezoelectric and magnetostrictive phases with mismatch in coefficient of thermal expansion and lattice at high temperature induces strain in the sintered composite. Thus, a post sintering heat treatment is necessary to homogenize the matrix grain structure, reduce the strain and remove the chemical or stress gradients at the interface. It has been shown that piezoelectric and dielectric properties of sintered ceramics improve after annealing. Annealing also enhances the magnetostrictive properties of some common ferrites such as CoFe2O4. The annealing and aging technique for this reason has been used in fabricating materials with strong permanent magnetism [25, 26]. In this case, precipitates with soft magnetic nature are dispersed in a hard magnetic matrix, resulting in one of the best hard magnets.

In this chapter, the percentage of the Ni ferrite (doped with Mn) was varied from 3 to 10% (by mole). Two different sintering temperature (1100° and 1125°C) was investigated to see the sintering behavior and the effect of ferrite percentage and sintering temperature on physical, piezoelectric and magnetoelectric properties of the particulate composite. This chapter investigates a new post sintering treatment such as the annealing and aging technique for synthesizing ME composites with the objective of achieving a strong coupling between ferroelectric and magnetic order parameters.

2. Experimental

2.1 Processing

Powders of PbO, ZrO_2 and TiO_2 (Alfa Aesar, Co. MA. USA) were mixed with alcohol and grinding media of YTZ (ϕ 5 mm, Tosoh Co. Tokyo, Japan) in a polyethylene jar for 24 hours. Similarly, NiO_2, Fe_2O_3, $MnCO_3$ was mixed and ball milled in same the fashion for 24 hours. After ball milling the powders were dried in an oven at 80°C. Then the powders were calcined (PZT at 750°C for 2 hours and NF at 1000°C for 5 hours). After calcinations, the powders were crushed and were examined by XRD to confirm the perovskite (for PZT) and spinel (for NF) phase. Then the powders were further crushed and sieved (US mesh # 270) very fine. Then the PZT and the NF powders were mixed stoichiometrically (for 3%, 5% and 10% NF by mole) and ball milled with alcohol and grinding media for 24 hours. After ball milling the powders were dried at 80°C, crushed in a mortar and sieved in a stainless-steel sieve of #170 US mesh. Then the powders were pressed to pellets of ϕ12.7x 1.5 mm in a hardened steel die using a hydraulic press under a pressure of 15 MPa. Pressureless sintering of composites was performed with Lindberg BlueM furnace with platinum foil (0.003 in. thick) at the bottom. The sintering temperatures were 1100° and 1125°C for 2 hrs. After sintering, samples were annealed at 800°C for 10 hours followed by air cooling and then aged at 400°C for 5 hours.

2.2 Characterization

The density of the samples was measured by Archimedes principle. XRD was perfromed in a Siemens Krystalloflex 810 D500 x-ray diffractometer on samples after calcination, sintering and post sintering treatment. For the measurement of % Spinel phase, the area under the curve for the perovskite (101) peak and spinel (311) peak was measured. Then the % Spinel was found taking the percentage of the area under spinel (311) peak among the total are under these two peaks. Specimens were polished with 0.3 μm powder, thermally etched and examined under Scanning Electron Microscope (Zeiss Leo Smart SEM). The annealed and aged specimens were chemically etched in a solution of 95% H_2O, 4% HCL and 1% HF. The average grain size of the composite was determined from SEM micrograph by linear intercept. X-ray mapping was done using the same Scanning Electron Microscope. TEM of sintered samples was done by JEOL – 1200 EX Scanning Transmission Electron Microscope.

2.3 Property measurement

An Ag-Pd electrode is painted on both surfaces of the pellets and fired at 825°C for 1 hr. The electroded specimens were polled by applying a D.C. field of 2.5 kV/mm for 20 minutes in a silicone oil bath at 120°C. The piezoelectric and dielectric properties were measured by APC YE 2730A d_{33} meter and an impedance analyzer (HP 4192, Hewlett Packard Co. USA). The Curie temperature was measured from Capacitance vs. Temperature graph with the help of Multi frequency LCR meter (HP 4274 A Hewlett Packard Co. USA). The magnetoelectric property was measured in terms of variation of dE/dH coefficient with an A.C. Magnetic field and no D.C. magnetic bias. The coefficient was measured directly as response of the sample to an A.C. magnetic input signal at 3 Hz and 60 Oe amplitude. The voltage generated from the composite was measured using a charge amplifier. The output signal from the amplifier was measured with an Oscilloscope (54601A, HP Co. USA). The output voltage is converted into dielectric displacement ($D = CV/A$), which can also be expressed in electric field ($E = D/\varepsilon_0\varepsilon$) and then the electric field divided by A.C. magnetic field gives the magnetoelectric voltage coefficient for PZT-NF composite.

3. Results and discussion

Figure 1 shows the density variation of the piezoelectric/magnetostrictive composites sintered at 1100 and 1125°C. Both shows similar pattern. The density

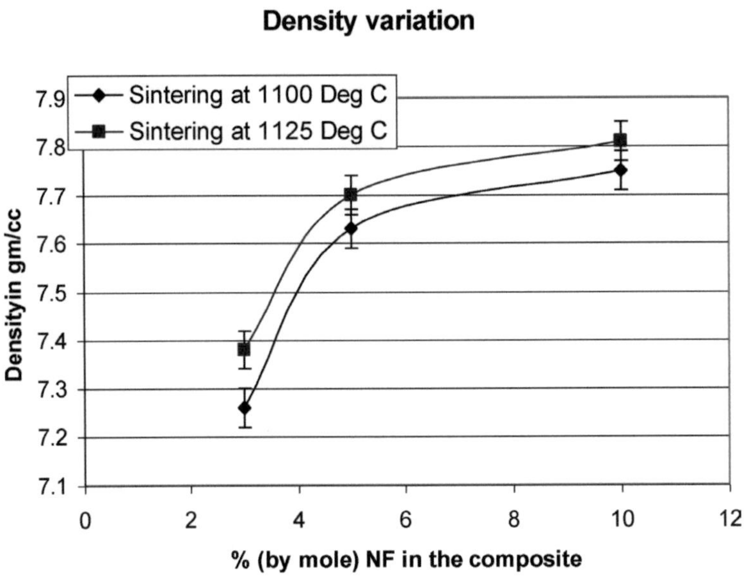

Figure 1.
Effect of % NFM on density at two different temperatures.

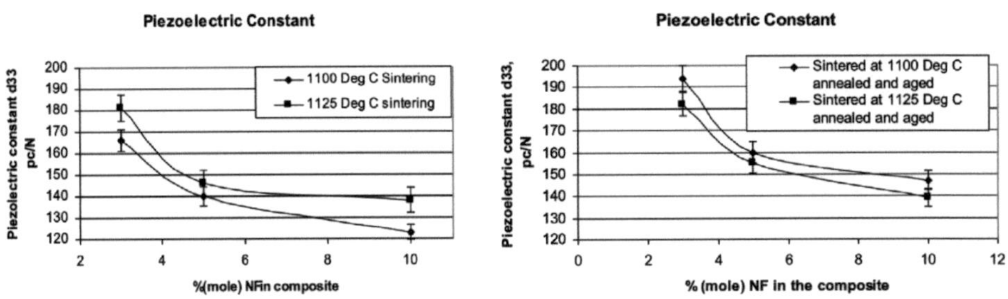

Figure 2.
Piezoelectric properties at different % NFM of sintered as well as annealed and aged samples.

increases with increasing percentage of NF and starts to plateau after 5 moles% of NFM. The samples sintered at 1125°C, showed higher densities than that of 1100°C. This is due to the better sintering behavior at higher (1125°C) temperature. **Figure 2(a)** shows the variation of the piezoelectric constant with NFM mole%. As the perovskite phase decreases with increasing NFM concentration, d33 starts to decrease. From **Figure 2(b)** an increase in d33 after annealing and aging was observed for both the sintering conditions. For the PZT-5NFM, 14% and 6% increase in d_{33} were recorded for samples sintered at 1100°C and 1125°C respectively. The difference in the magnitude of d_{33} before and after annealing and aging for other two composite samples (3% and 10% NFM) sintered at 1125°C was small. This can be associated to concurrent rise in the internal strain due to structural dissimilarity and domain size. Increase in stress decreases d33 while increase in domain size increases d_{33}. **Figure 2(c)** shows the variation of the dielectric constant with NFM concentration. The dielectric constant magnitude for the samples sintered at 1100°C showed a decrease of about 10% after thermal treatment. Based on the behavior of the piezoelectric and dielectric constant, it can be seen that the total strain magnitude remains of the same order for the samples before and after the thermal treatment.

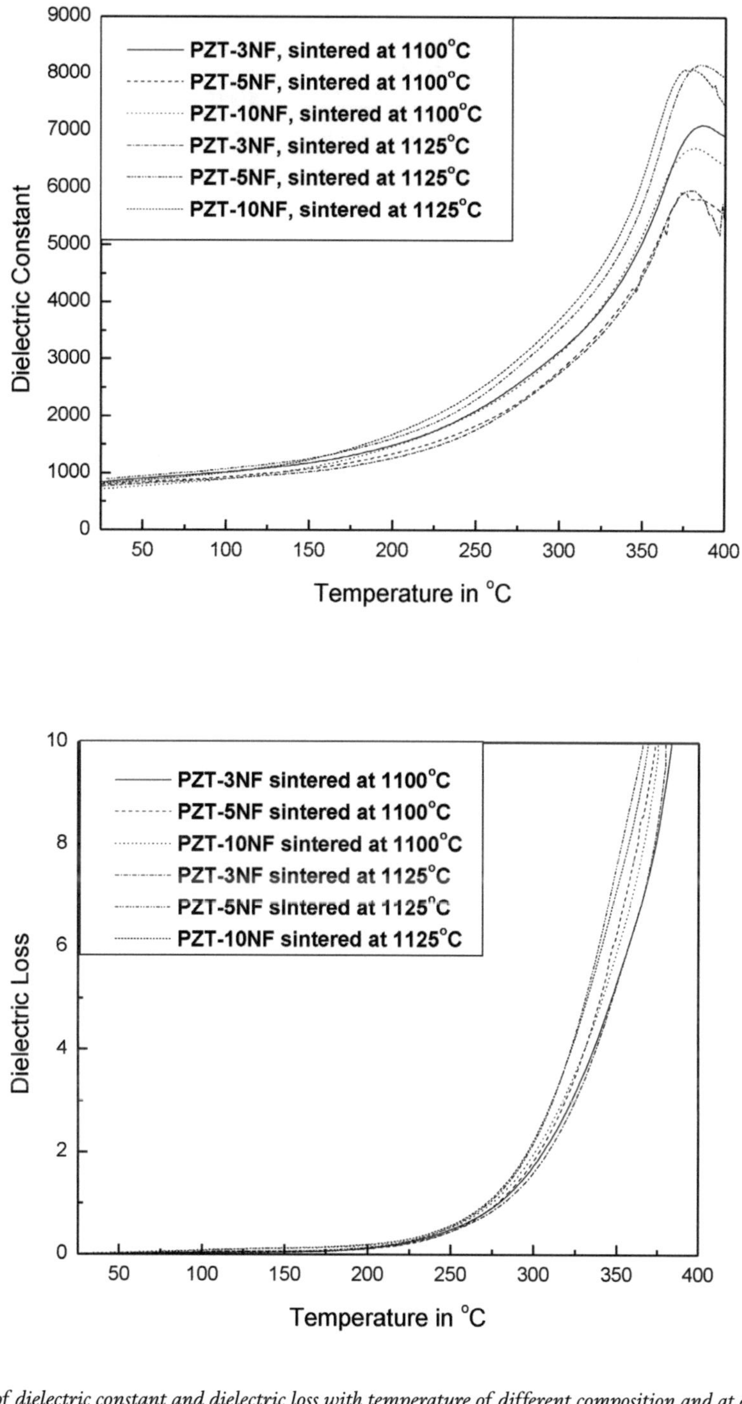

Figure 3.
Variation of dielectric constant and dielectric loss with temperature of different composition and at different sintering temperature [27–30].

Figure 3(a) and **(b)** shows the dielectric constant and loss factor as a function of temperature. The dielectric properties were measured at 1 kHz under 1 V excitation. The Curie temperature (T_{max}) measured for all the samples compositions were in the vicinity of 375°C with slight decrease in T_{max} with increasing % NFM. The loss factor (tanδ) measured was ~2% below 80°C. Beyond 200°C it increases sharply. The space charge effect and low resistivity of the PZT-NFM samples at elevated

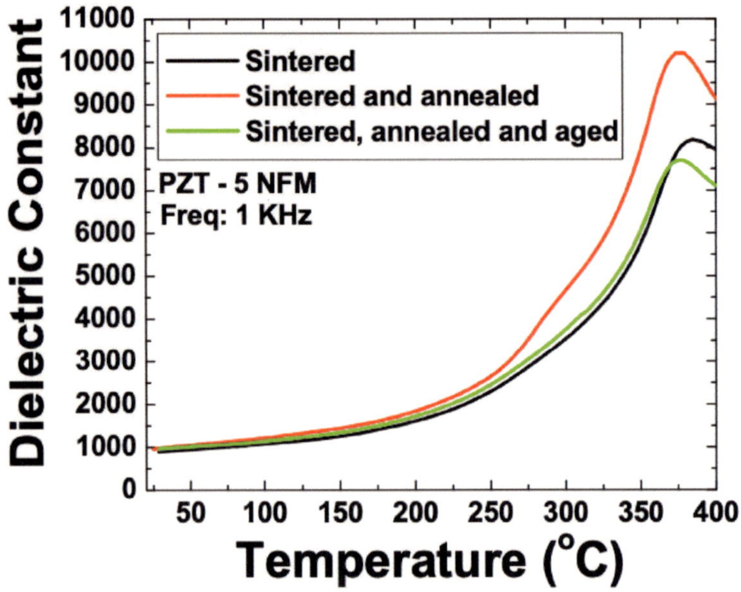

Figure 4.
Dielectric constant as a function of the temperature at 1 kHz for PZT-5 NFM at three different conditions [28, 29].

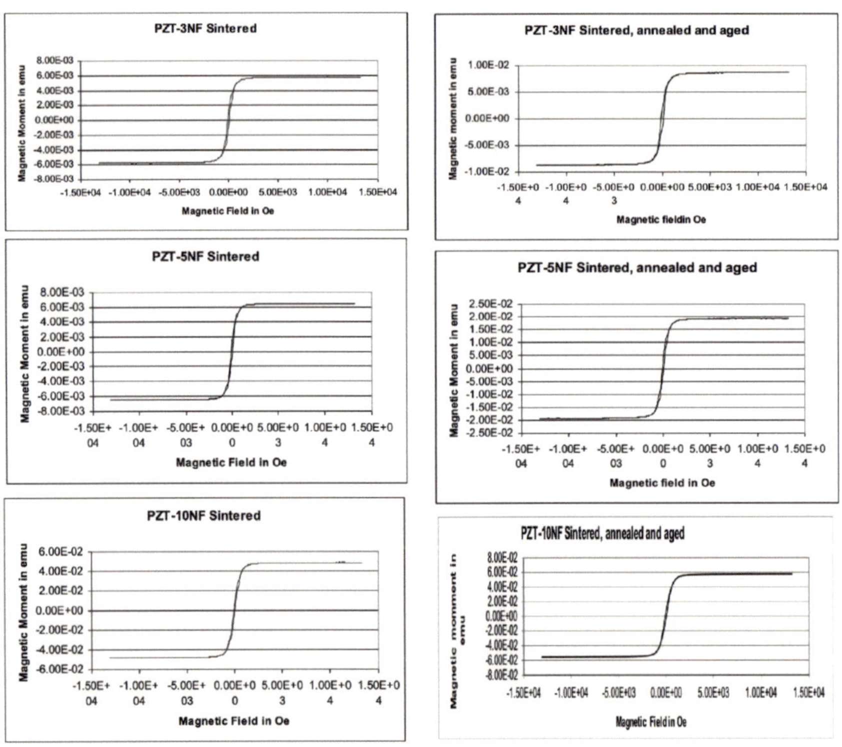

Figure 5.
Magnetic field H vs magnetic moment curve of different composite.

temperature caused the rapid rise of dielectric loss around curie temperature. The T_{max} did not change much regardless of the thermal treatment. The dielectric constant vs. temperature for PZT-5 at % NFM composites before and after different

	PZT-3NF			PZT-5NF			PZT-10NF		
	Powder	Sintered at 1100°C	Sintered and annealed	Powder	Sintered at 1100°C	Sintered and annealed	Powder	Sintered at 1100°C	Sintered and annealed
H_c (Oe)	124	86.94	89.91	97.39	75.62	78.76	109.9	69.78	70.24
M_r (μemu)	195	1010	1480	270.8	904.1	2538	219.1	5305	5575
M_s (m emu)	0.817	5.766	8.653	1.633	6.472	19.25	1.187	48.64	57.13
M_r/M_s	0.2356	0.1751	0.1711	0.1658	0.1397	0.1318	0.1846	0.1091	.0976
S	0.00575	0.005	0.0043	0.0047	0.0048	0.0036	0.0057	0.0034	0.0029
Resonance (Hz)	525	509.4	465.2	612.2	543	463	592	453.9	439.8

Table 1.
Magnetic properties of different PZT-NF powders [28].

thermal conditions were shown in **Figure 4**. The ferroelectric Curie temperature only decreased slightly (~8°C) by annealing and aging. This drop in T_{max} can be attributed to the stress relaxation which can be precisely understand by diffraction studies. During annealing as the composites were soaked in 800°C for 10 hours, grain grown occurs which can result into increased dielectric constant.

Figure 5(a) to **(f)** shows the variation of magnetic properties as a function of thermal treatment for the different composites. A substantial enrichment in remanent (M_r) and saturation (M_s) magnetization was observed after the thermal (annealing and aging) treatment. **Table 1** shows the magnetic data for the calcined powder, sintered samples and thermally treated samples. It is clearly observed that the magnetization values increased sharply after thermal treatment. No difference in coercive fields were observed between the sintered and thermally treated samples whereas remarkable differences were observed in magnetic resonance which decreased significantly after thermal treatment. These results can be explained if it is assumed that the size of spinel phase increases with the thermal treatment. As there are basic resemblance in the oxygen synchronization chemistry between the perovskite and spinel structure, it leads to the lattice dimensions that are compatible with the spinel building blocks (considering growth is along the c-axis).

Figure 6(a) shows the X-Ray diffraction patterns of composite sintered at 1125°C for 2 hrs. A pure perovskite phase was obtained with small fraction of 311 peak of spinel phase only observed for PZT-10 NFM. No other phase was detected. PZT composition of Zr: Ti = 52:48 was selected as it is closer to the morphotropic phase boundary (MPB) providing high piezoelectric property. On modification with NFM the perovskite phase was found to exhibit rhombohedral symmetry as shown by reduced splitting of 200/002 peaks. As expected, a higher sintering temperature resulted in higher content of spinel phase. **Figure 6(b)** shows the XRD pattern for same composition of samples after post sintering thermal treatment. In all three diffraction patterns, an increase in the fraction of spinel phase were clearly observed. The fraction of the spinel phase present was computed using the expression:

$$\%Spinel = \frac{Area\ under\ the\ spinel\ peak\,(311)}{Area\ under\ the\ perovskite\ peak\,\{(110)+(101)\} + Area\ under\ the\ spinel\ peak\,(311)} x\,100 \quad (1)$$

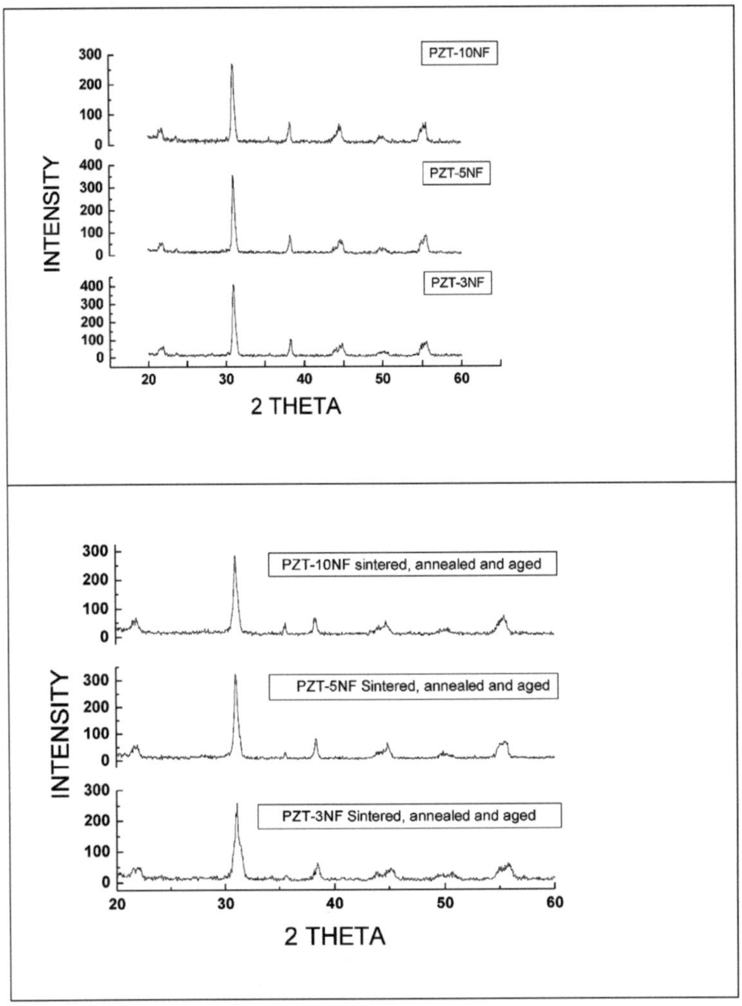

Figure 6.
XRD pattern of different compositions: a. after sintering, b. after annealing and aging.

PZT-10 NFM sintered at 1125°C showed an increase of % spinel of 1.03% (from 6.82% after sintering to 7.85% after annealing and aging). This is a significant increase considering the dissimilarity in the lattices of two phases. Comparing this result with the annealing and aging treatment, it can be theorized that there is a likelihood of homogenization in the PZT-NFM system.

Figure 7**(a)** and **(b)** shows the Magnetoelectric coefficient for 3, 5 and 10 mole% of NFM. Data are given in each figure for the three different thermal histories. For PZT-10NFM, the ME coefficient increased from 60 mV/cm-Oe after sintering to 88 mV/cm-Oe after annealing and aging which is nearly 50% increase. This is due to the reduction in misfit strain between perovskite and spinel phase, decrease in interface micro such as porosities and cracks after annealing. Reducing the interface defects, would increase the ability of piezoelectric domains to elastically react to strains induced on it by bordering magnetostrictive phases, or vice versa. To achieve high ME properties, the boundary conditions between phases needs to be as mechanically free as possible. Enhancements in the ME coefficient may also come from increased magnetization from Mn^{+3} to Mn^{+2} conversion, as an enhanced magnetic permeability has been reported to increase the effective piezomagnetic coefficient ($d\lambda/dH$).

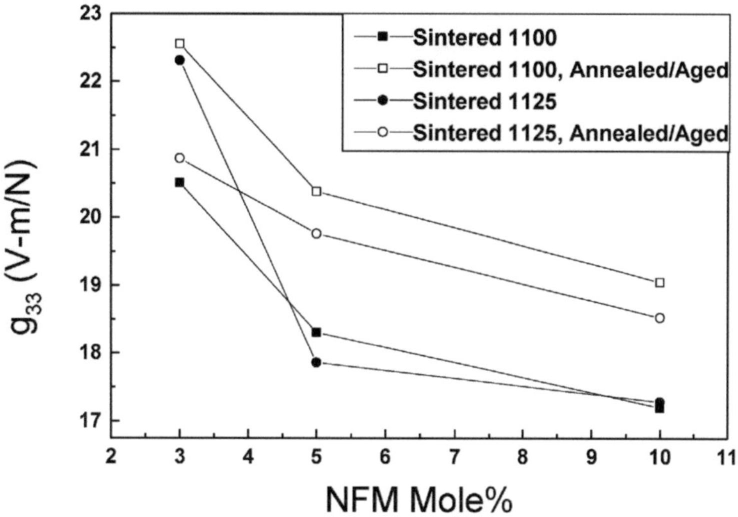

Figure 7.
Variation of magnetoelectric coefficient as a function of % NFM [27–29].

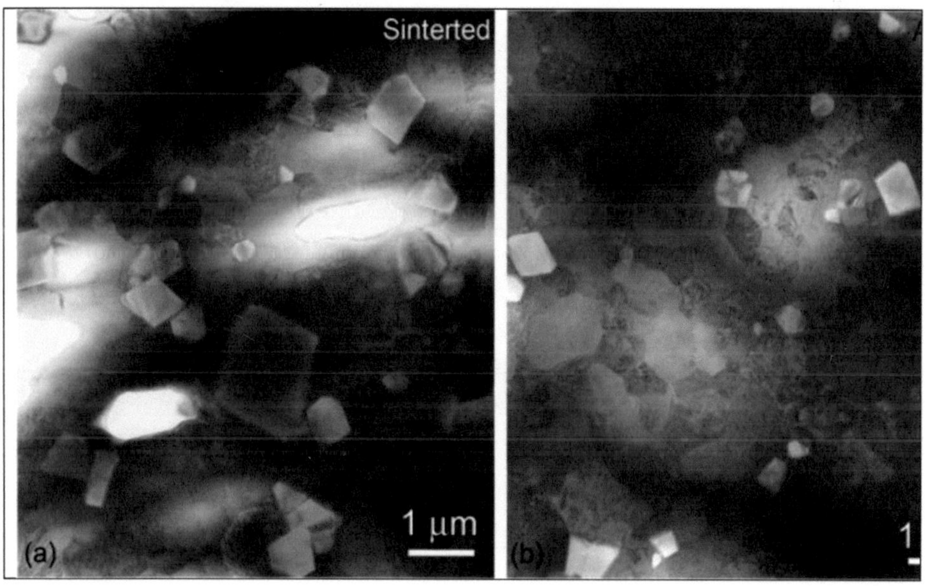

Figure 8.
Bright field TEM images PZT-5 NFM composites after (a) sintering, and (b) annealing [30].

Figure 8(a) shows Transmission Electron Microscopy image of as-sintered sample. This image consists of facet phases (bright contrast) NFM particles embedded in the PZT matrix. The NFM particles vary from 300 nm to 1500 nm. **Figure 8(b)** shows TEM images of annealed and aged samples. Few distinctions of microstructural characteristics are observed from those shown in **Figure 8(a)**. The density of the NFM particles in this annealed sample is much less than that in the sintered sample. In addition, the NFM particles in the annealed sample have a typical size of 500 nm, much smaller than that in the sintered sample.

High magnification bright field TEM images of the sintered, annealed and aged samples were shown in **Figure 9(a)–(c)**. Misfit strain fields close to the

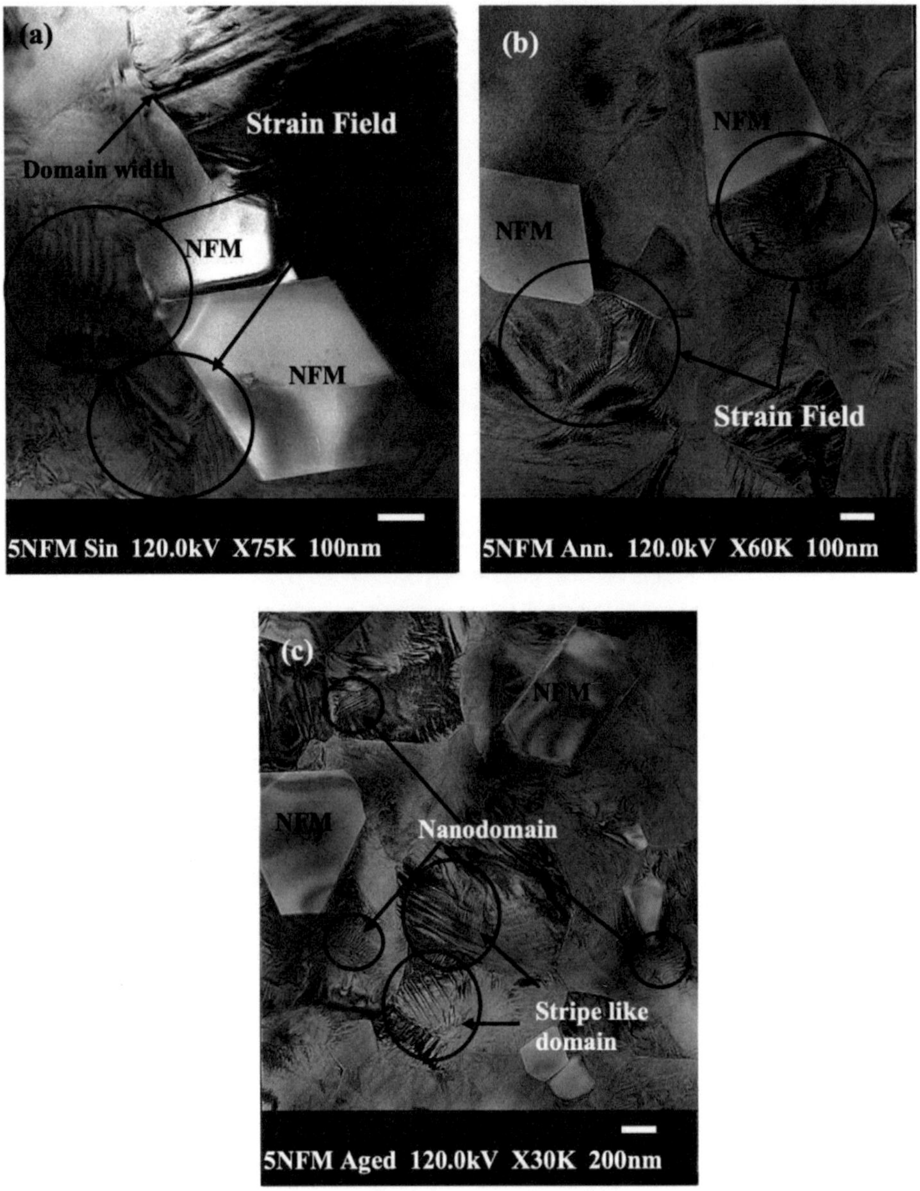

Figure 9.
Bright field TEM images PZT-5 NFM composites after (a) sintering, (b) annealing and (c) aging [30].

PZT/NFM were observed. These strain fields developed at the interface to acclimatize the mismatch in the PZT and NFM lattice. The domain patterns had larger width, which is the characteristic of 90° domains and there is intergranular discrepancy in domain width [31]. The image after annealing shows reduced misfit strain near the PZT/NFM interface. The remnant strain fields after annealing consist of additional constituent due to quenching process after annealing. Diffused grain boundaries were observed in low magnification images. The image after the aging in **Figure 9(c)** shows noticeably reduced strains. There are stripe like morphologies and they are extended from grain boundary to grain boundary. A finer scale domain structure is also observed to exist within larger domain patterns. This finer domain pattern has striation like morphology and is periodically spaced [32, 33].

4. Conclusion

In the chapter the effect of Mn doped Nickel ferrite percentage in Lead Zirconate Titanate (PZT) at different sintering temperature on the piezoelectric and magnetoelectric behavior was tried to find out and also a new post sintering treatment (annealing and aging) is introduced to enhance the magnetoelectric property. It was found that with increasing percentage of NF the piezoelectric constants, dielectric constant decreases and dielectric loss increases. The starting powder of PZT-NF (3%, 5% and 10% NF) was found ferromagnetic where the coercive field varies from 97 to 124 Oe. Post sintering treatment like annealing and aging increases the saturation magnetization M_s from 48 to 57 memu for PZT-10NF. High values (~144 mV/cm Oe) of magnetoelectric coefficient were found in 10% NF, sintered at 1125°C. The preferential distribution of NF particles in large grains of PZT matrix was found in X-ray mapping. In order to even out the distribution of NF particles and increase the magnetic phases, annealing followed by aging was introduced, as annealing is a homogenization treatment. Due to annealing the piezoelectric property increases as well as magnetoelectric coefficient. The magnetoelectric coefficient was found 211 mV/cm Oe at 3 Hz with no D.C. magnetic field bias. This is higher than any values so far reported in PZT-10% NF and sintering temperature at 1125°C. A supersaturated structure of NF in PZT is formed by annealing at high temperature followed by faster cooling. Aging allows the NF particles to distribute evenly in the structure as a different phase. The XRD pattern also shows the introduction of spinel phases in PZT matrix. The SEM images also support the grain homogenization treatment. The introduction of this post sintering treatment (annealing and aging) enhanced the property and can be further improved with optimum parameters of the treatment.

Acknowledgements

The authors are sincerely thankful to Prof. Shashank Priya (Penn State University) for technical guidance and Prof. J. Ping Liu's research group (University of Texas at Arlington) for their help in the magnetic characterization.

Author details

Rashed Adnan Islam
Product Integrity Engineering – Google Devices and Services, Google LLC, Mountain View, CA, United States

*Address all correspondence to: rashed@google.com

References

[1] Wang J, Zheng H, Lofland SE, Ma Z, Ardabili LM, Zhao T, Riba LS, Shinde SR, Ogale SB, Bai F, Viehland D, Jia Y, Schlom DG, Wuttig M, Roytburd A, and Ramesh R (2004) Science. 303: 661.

[2] Ederer C and Spaldin N (2004) Nature Mater. 3: 849.

[3] Eerenstein W, Mathur ND, and Scott JF (2006) Nature. 442: 759.

[4] Hur N, Park S, Sharma PA, Ahn JS, Guha S and Cheong SW (2004) Nature. 429: 392.

[5] Astrov DN (1960) Sov. Phys. JETP. 11: 708.

[6] Wang J, Neaton JB, Zheng H, Nagarajan V, Ogale SB, Liu B, Viehland D, Vathiyanathan V, Schlom DG, Waghmare UV, Spaldin NA, Rabe KM, Wuttig M, and Ramesh R (2003) Science. 299: 1719.

[7] Van Aken BB, Palstra TTA, Filippetti A, and Spaldin NA (2004) Nat. Mater. 3: 164.

[8] Zheng M, Wan JG, Wang Y, Yu H, Liu JM, Jiang XP and Nan CW (2004) J. Appl. Phys. 95 (12): 8069.

[9] Ryu J, Carazo AV, Uchino K, and Kim H (2001) J. Electoceram. 7: 17.

[10] Flores VC, Baques DB, Flores DC, and Aquino JAM (2006) J. Appl. Phys. 99: 08J503.

[11] Van Suchetelene J (1972) Philips Research Report. 27: 28.

[12] Weng L, Fu Y, Song S, Tang J, and Li J (2007) Scrip. Mater. 56: 465.

[13] Ren SQ, Weng LQ, Song SH, Li F, Wan JG, and Zeng M (2005) J. Mater. Sci. 40 (16): 4375.

[14] Wu D, Gong W, Deng H, and Li M (2007) J. Phys. D: Appl. Phys. 40: 5002.

[15] Srinivasan G, DeVreugd CP, Flattery CS, Laletsin VM, and Paddubnaya N (2004) Appl. Phys. Lett. 85 (13): 2550.

[16] Ryu J, Priya S, Uchino K, Viehland D, and Kim H (2002) J. Kor. Ceram. Soc. 39: 813.

[17] Ryu J, Priya S, Uchino K, and Kim H (2002) J. Electroceram. 8: 107.

[18] Srinivasan G, Rasmussen E, Levin B, and Hayes R (2002) Phys. Rev. B 65: 134402.

[19] Dong SX, Zhai J, Li JF, and Viehland D (2006) J. Appl. Phys. 88: 082907.

[20] Dong S, Zhai J, Li JF, and Viehland D (2006) Appl. Phys. Lett. 89: 252904.

[21] J. Ryu, A.V. Carazo, K. Uchino, and H. E. Kim. Journal of Electroceramics, Vol. 7, 2001, pp. 17-24

[22] T.G. Lupeiko, S.S. Lopatin, I.V. Lisnevskaya, and B.I Zvyaginstev. Inorganic Materials Vol. 30, 1994, pp 1353-1356.

[23] C.W. Nan. Physical Review Vol. 50 (9), 1994, pp. 6082-6088.

[24] Y. X. Liu, J.G. Wan, J.M. Liu, and C. W. Nan. J. Appl. Phys. Vol. 94 (8), 2003, pp. 5118-5122.

[25] M.K. Griffiths, J.E.L. Bishop, W.J. Tucker, and H.A. Davies. J. Magnet. Magnetic Mater. 234, 331-352 (2001).

[26] Z. Sufen, J. Hanmin, W. Xuefeng, and Y. Yu. J. Magnet. Magnetic Mater. 247 [1], 15-18 (2002).

[27] R. A. Islam and S. Priya. Integ. Ferroelec. 82 (1), 1 (2006).

[28] R. A. Islam and S. Priya. Jap. J. Appl. Phys. 45 (5), L128–L131 (2006).

[29] R.A. Islam and S. Priya. J. Appl. Cer. Tech. 3(5), 353 (2006).

[30] R.A. Islam. Composition – Microstructure – Property Relationships in Dual Phase Bulk Magnetoelectric Composite [Thesis], University of Texas at Arlington, 2008.

[31] J.F. Li, X. Dai, A. Chow, and D. Viehland. J. Mater. Res. 4, 926 (1995).

[32] X. Bi, S. Chu, J. G. Zhu, and D. Laughlin. J. Appl. Phys. 99, 08B306 (2006).

[33] Z. Xu, X. Dai, D. Viehland, and D. Payne. J. Am. Ceram. Soc. 78 [8], 2220 (1999).

Section 2

Applications

Chapter 4

Application of Ferrites as Electrodes for Supercapacitor

Ankur Soam

Abstract

Apart from the magnetic properties, ferrites have been considered as efficient electrodes for next generation energy storage devices. This chapter will include applications of spinel ferrites such as $MnFe_2O_4$, $CoFe_2O_4$, $ZnFe_2O_4$ and $NiFe_2O_4$ in supercapacitor. In ferrites, the charge storage arises from the fast-reversible surface redox reactions at the electrode/electrolyte interface. In particular, the electrode material with high specific capacitance, wide range of operating potential, low synthesis cost and its availability on the earth are highly desirable to fabricate a supercapacitor. Ferrites with mixed oxidation states have proved as promising electrodes in supercapacitors. In this chapter, we summarize the different synthesis methods of ferrites based nanocomposites and their electrochemical properties for supercapacitor application.

Keywords: ferrites, nanocomposites, electrochemical properties, electrodes, supercapacitor

1. Introduction

The continuous depletion and consequently the increased cost of the fossil fuel has now become an economic problem for a nation. Moreover, the production of CO_2 from massive use of fossil fuel in transportation and industrial operations increases the greenhouse gases which are responsible for significance change in climate (global warming). In future, the demand of fossil fuel is expected to be increased rapidly. Therefore, some alternative energy storage systems need to be developed in order to meet the demand of energy consumption. Battery is being widely utilized in electric vehicles and electronic devices because of its large energy density [1–4]. However, the maintenance at regular interval and low power density are some drawbacks with battery.

Among various energy storage devices, supercapacitor technology has attracted tremendous attention to be used in high power application because of their higher power density and longer cycling life [5–9]. **Figure 1** depicts the power density and energy density of capacitor, battery and supercapacitor. The capacitor with largest power density occupies the top position, however, the energy density is much lower. Battery can exhibit larger energy density but with lower power density. The supercapacitors occupy the important space between capacitor and battery with larger power density than batteries and greater energy density than capacitors. Supercapacitors are considered suitable candidates as energy storage in portable consumer electronic devices, memory back-up systems, microelectro-mechanical systems, hybrid electric vehicles and medical devices [10–17]. Further

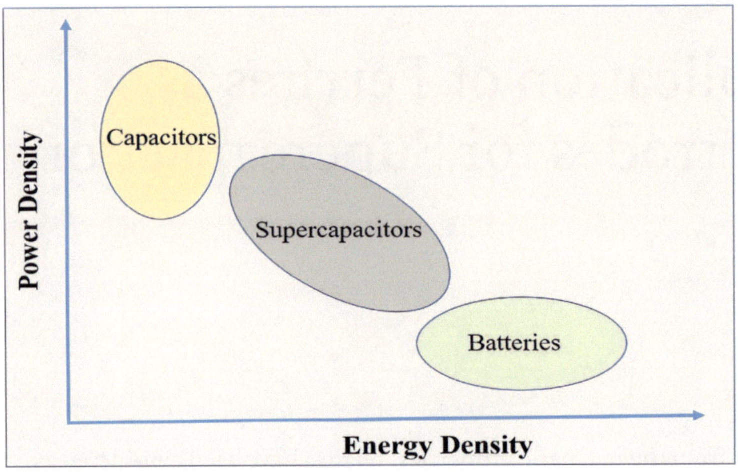

Figure 1.
Ragone plot for various charge storage devices. Supercapacitors occupy the space between capacitors and batteries with larger energy density than capacitors and larger power density than batteries.

improvements are being made in order to extend the applications of supercapacitor in different purpose [18–20]. Due to a simple structure (similar electrodes), supercapacitor technology can be integrated on a Si chip with energy harvesters [21–23].

Supercapacitor can store an excess energy from the harvester and return back when required. The supercapacitor performance is governed by the electrodes, current collectors, separator and electrolyte. The surface morphology and electrical properties of electrodes are the major factors which mainly control the energy storage in supercapacitor. In this connection, a lot of efforts are being made towards developing new materials for electrodes and improving their electrochemical properties [24–29]. Many materials and their composites have been explored as electrodes for supercapacitor [30, 31].

Supercapacitor electrodes can be categorized in two types, 1) Metal oxides, which involve faradaic process to store the charge (Pseudocapacitor) [32–34], 2) carbon and silicon based materials, these materials store the charge in electric double layer (EDL) [27, 35–37]. EDLCs exhibit high rate capability and longer cycle life, but low energy storage capacity is a major issue for them [38–40]. On the other hand, metal oxide based supercapacitor exhibits larger capacitance and energy density than EDLC but the poor rate capability and limited charging/discharging cycle numbers are some of their drawbacks [7, 15, 41]. To design a supercapacitor with larger energy density without compromising the rate capability is a major challenge. In this context, several electrode materials and their combinations have been evaluated for high performance supercapacitor [5, 42–44]. The ferrite materials are also being considered as potential electrodes in supercapacitor because of their different oxidation states, low price, environmental benignity, and their large abundance [26, 45–49]. Moreover, their synthesis process is simple and suitable for production at industrial scale. MFe_2O_4 (M = Mn, Co, Ni, Zn, or Mg) ferrites have been extensively used in supercapacitor. These binary oxides can offer large capacitance due to involvement of two ions in redox reactions [13, 47, 50]. Subsequently, several studies were performed on ferrite materials such as nickel ferrite, bismuth ferrite, cobalt ferrite, manganese ferrite, as electrodes in supercapacitor [46, 51–53]. Ferrites of the form MFe_2O_4 (M = Ni, Co, Zn, etc.) have been considered as potential electrodes in energy storage devices because of their good chemically stability and electronic properties. Moreover, their nanocomposite can be synthesized using water based solution without any organic solvent (**Figure 2**). In this chapter, research progress

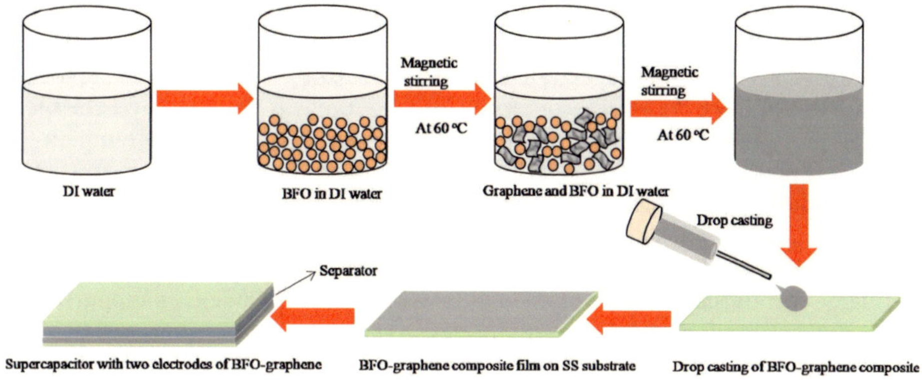

Figure 2.
Synthesis of ferrite (bismuth ferrite)–graphene nanocomposite as electrodes for supercapacitor using water based solution [34]. Bismuth ferrite and graphene were mixed in DI water and then deposited on a conducting substrate by drop casting process to fabricate the electrodes for supercapacitor.

on ferrite based electrodes (MFe_2O_4 types) for supercapacitor have been summarized. The supercapacitor performance of ferrite depends on their morphology, synthesis process, used precursors, and composition.

2. Cobalt ferrite ($CoFe_2O_4$)

$CoFe_2O_4$ with high magnetic properties has good mechanical hardness and chemical stability, which make it suitable candidate in magnetic device applications [54]. Apart from above properties, $CoFe_2O_4$ has also shown good electrochemical performance in supercapacitor [52, 55–59]. $CoFe_2O_4$ nano-flakes synthesized by chemical bath deposition process have been utilized in supercapacitor [52]. The nano flakes were deposited on a stainless steel substrate with thickness of 5.3–7.0 μm. The electrochemical properties were investigated in three electrode system with 1 M NaOH electrolyte. The nano-flakes showed a specific capacitance of 366 F g^{-1} (interfacial capacitance of 0.110 F cm^{-2}) at scan rate of 5 mVs^{-1}. The nano-flakes electrode could preserve 190 Fg^{-1} capacitance at scan rate of 100 mV/s^{-1}. At such larger scan rates, the inner active sites of the electrodes might not participate in redox process, resulting in decrease in the capacitance. The $CoFe_2O_4$ nano-flakes retained 90.6% capacitance after 1000 cycles.

Pawar et al. [60], have synthesized $CoFe_2O_4$ nanoparticles (average size 23 nm) by sol–gel method for supercapacitor application. $CoFe_2O_4$ nanoparticles exhibited pseudocapacitive behavior in 1 M KOH electrolyte with a three-electrode system as observed from the CV curves. The specific capacitance determined from galvanostatic charge discharge process were found to be 15 Fg^{-1} at current density of 0.6 Ag^{-1}. This lower value of the capacitance may be due to the poor electrical conductivity of $CoFe_2O_4$ nanoparticles. An improved electrochemical performance was achieved with $CoFe_2O_4$/FeOOH nanocomposite synthesized by one-step hydrothermal approach [61]. A mixture of $CoFe_2O_4$/FeOOH nanocomposites, carbon black and polyvinylidenefluoride (PVDF) in ratio of 80:10:10 wt% was used as electrode. The electrode of $CoFe_2O_4$ with FeOOH exhibited good value of capacitance at larger currents which is essential for a supercapacitor. The electrode showed specific capacitance of 332.4, 319.4, 257.2, 239, 193.1, and 180 F g^{-1} at the current densities of 0.5, 1, 2, 5, 8, and 10 A g^{-1}, respectively. About 8.7% loss in capacitance was observed after 1000 cycles.

Mesoporous $CoFe_2O_4$ thin film also demonstrated good value of specific capacitance of 369 Fg^{-1} at 2 mV/s^{-1} in 1 M KOH electrolyte with wide potential window of −1.2 to +0.5 V [62]. The capacitance degraded to 167 Fg^{-1} upon increasing the scan rate to 100 mV/s^{-1} . The mesoporous $CoFe_2O_4$ film was prepared by a chemical spray pyrolysis technique from the aqueous medium at 475°C substrate temperature. The film was observed to be uniform on the substrate and free from any crack with mesoporous type surface morphology. $CoFe_2O_4$ thin film consists of grain with size in the range of nm. The electrode of mesoporous $CoFe_2O_4$ film showed power density of 28.74 $kWkg^{-1}$ with maintaining energy density of 27.14 $Whkg^{-1}$.

A large specific capacitance of 429 Fg^{-1} was obtained by $CoFe_2O_4$ nanoparticles in 6 M KOH electrolyte at 0.5 Ag^{-1} [63]. These nanoparticles were prepared by hydrothermal and coprecipitation methods using nitrates, chlorides and acetates precursors with average size from 11 to 26 nm and surface area of ~34 $m^2\ g^{-1}$. The nanoparticles has shown an excellent capacitance retention of 98.8% after 6000 cycles at high current density of 10 Ag^{-1}. The above results indicate that $CoFe_2O_4$ nanoparticles with the above morphology may be a promising electrode material for supercapacitor. A composite of reduced graphene oxide and $CoFe_2O_4$ (RGO–$CoFe_2O_4$) was examined for supercapacitor application [64]. The electrode of RGO–$CoFe_2O_4$ showed a specific capacitance of 123.2 F g^{-1} which is larger than that of individual constituents RGO (89.9 F g^{-1}) and $CoFe_2O_4$ (18.7 F g^{-1}) at current density of 5 mA cm^{-2}. However, about 22% loss in capacitance was observed for the electrode RGO–$CoFe_2O_4$ after 1000 cycles.

Xiong et al. have developed ternary nanocomposite of cobalt ferrite/graphene/polyaniline as electrode for high-performance supercapacitor [65]. Hydrothermal method was used to make $CoFe_2O_4$ nanoparticles and graphene nanosheets and then polyaniline (PANI) coating was performed on $CoFe_2O_4$ by in situ polymerization process. A large specific capacitance of 1133.3 F g^{-1} at a scan rate of 1 mVs^{-1} was observed by the hybrid ternary nanocomposite electrode in 1 M KOH electrolyte. 716.4 F g^{-1} specific capacitance was determined in two electrode system in the same electrolyte and at the same scan rate of 1 mVs^{-1}. The electrode demonstrated long cycle stability about 96% retention of initial capacitance after 5000 cycles. The synergistic effects of three components in the ternary composite improved the electrochemical performance of the electrode. The graphene nanosheets greatly enhance the electron transfer in the electrode and surface area of the electrode, leading to increase in the overall capacitance [8]. In addition, PANI also contributes to the pseudocapacitance of $CoFe_2O_4$ nanoparticles.

Cobalt ferrite nanoparticles were also used as negative electrode in an asymmetric supercapacitor with positive electrode of $Co(OH)_2$ and $Co_2Fe(CN)_6$ particles [55]. The negative electrode of $CoFe_2O_4$ showed a specific capacitance of 758.86 F g^{-1} at 2 mV s^{-1} in 1 M KOH electrolyte. Overall, the asymmetric supercapacitor with electrode combination of $CoFe_2O_4$‖AC, $CoFe_2O_4$‖$Co(OH)_2$ and $CoFe_2O_4$‖$Co_2Fe(CN)_6$ provided the specific capacitance of 339, 127 and 125 F g^{-1} at 1 mVs^{-1} scan rate.

3. Manganese ferrite ($MnFe_2O_4$)

$MnFe_2O_4$ based electrodes exhibited good capacitive properties in aqueous electrolyte [53, 66–70]. It is observed that $MnFe_2O_4$ stores the charge by pseudo mechanism in aqueous electrolyte [71]. Shin-Liang Kuo et al. [71] have shown that charge storage in $MnFe_2O_4$ involves insertion/extraction of proton into/from the lattice at both the Mn- and Fe-ion sites. In that study, $MnFe_2O_4$ was synthesized by coprecipitation method for supercapacitor application. For electrochemical characterization, the

$MnFe_2O_4$ powder was mixed carbon black and PVDF, and then coated on current collector. Electrochemical performance was determined with a three-electrode cell in 1 M KCl aqueous solution. The overall specific capacitance of 63.4 Fg^{-1} was determined for the electrode and specific capacitance of 115 Fg^{-1} for $MnFe_2O_4$ electrode. Baoyan Wang et al. [53] have studied the effect of surfactants on the electrochemical performances of $MnFe_2O_4$ synthesized by solvothermal method. The capacitive performances of $MnFe_2O_4$ colloidal nanocrystal cluster was observed to be larger than $MnFe_2O_4$ hollow sphere in aqueous $LiNO_3$ electrolyte. In that work, an almost rectangular CV curves were obtained for $MnFe_2O_4$ in a potential range of −0.4-1.5 V. Addition of surfactants leads to increase in the capacitance of $MnFe_2O_4$ in $LiNO_3$ electrolyte. It may be due to the reduction of interfacial tension between electrode and electrolyte in presence of surfactants with promoting the diffusion of lithium ions. After addition of different surfactants, SDS (Anionic surfactant sodium dodecyl sulphate), Triton-X-100 (non-ionic surfactant p-toctylophenol) and P123 (poly(ethylene glycol)-block-poly(propylene glycol)-blockPoly(ethylene glycol)), the capacitance increased about 36.8%, 22.8% and 12.8%, respectively.

V. Vignesh et al. [69] have reported electrochemical properties of $MnFe_2O_4$ spherical nanoparticles (20–50 nm) synthesized by simple and facile coprecipitation method. The capacitor performance was evaluated in different electrolytes, 1 M $LiNO_3$, 1 M Li_3PO_4 and KOH. The $MnFe_2O_4$ nanoparticles showed specific capacitance of 173, 31 and 430 F g^{-1} in electrolytes of 3.5 M KOH, 1 M $LiNO_3$ and 1 M Li_3PO_4, respectively. However, excellent rate performance was observed in 3.5 M KOH electrolyte with good retention of capacitance at higher current densities. Supercapacitor with two electrodes of $MnFe_2O_4$ nanoparticles exhibited specific capacitance of 245 F g^{-1}, and energy density and power density of 12.6 Wh kg^{-1} and 1207 W kg^{-1}, respectively in 3.5 M KOH electrolyte.

Further, the electrochemical performances of $MnFe_2O_4$ colloidal nanocrystal clusters (CNCs) was investigated in symmetric supercapacitors with different aqueous electrolytes [72]. The specific capacitances of $MnFe_2O_4$ electrode was found to be 97.1, 93.9, 74.2 and 47.4 F g^{-1} in electrolytes 2 M KOH, 2 M NaOH, 2 M LiOH and 2 M Na_2SO_4, respectively. It was found that $MnFe_2O_4$ CNCs exhibited better performance in 6 M KOH electrolyte with the specific capacitance of 152.5 F g^{-1} and retention of capacitance of about 76% after 2000 cycles. $MnFe_2O_4$ colloidal nanocrystal assemblies (CNAs) with size of 420 nm, composed of 16 nm nanoparticles showed specific capacitance of 88.4 Fg^{-1} calculated at the current density of 0.01 Ag^{-1} [70]. When the current increased from 0.01 to 2 Ag^{-1} $MnFe_2O_4$ CNAs retained 59.4% capacitance, and 69.2% capacitance after 2000 cycles. The electrochemical performance of $MnFe_2O_4$ CNAs was related to the size of primary nanoparticles in the CNAs.

Further improvement in $MnFe_2O_4$ based supercapacitor was made by making nanocomposite of $MnFe_2O_4$ with grapheme [66, 68, 73–75]. Isara Kotutha et al. [73] have used one-pot hydrothermal approach to prepare rGO/$MnFe_2O_4$ nanocomposite. A maximum specific capacitance of 276.9 Fg^{-1} was determined for the rGO/$MnFe_2O_4$ nanocomposite at scan rate of 10 mVs^{-1} in 6.0 M KOH electrolyte. A flexible supercapacitor of $MnFe_2O_4$/graphene using current collectors of flexible graphite sheets has been fabricated [66]. The flexible supercapacitor exhibited specific capacitance of 120 F g^{-1} at 0.1 A g^{-1} with retaining 105% capacitance after 5000 cycles.

Larissa H. Nonaka et al. [68] have achieved 195 Fg^{-1} capacitance for $MnFe_2O_4$ nanoparticles on a crumpled graphene sheet at scan rate of 0.5 Ag^{-1} in 0.05 M KCL electrolyte. A pseudocapacitive behavior was observed in the CV curves which indicates that there is major contribution from $MnFe_2O_4$ to the overall capacitance of the hybrid electrode. A larger specific capacitance of 454.8 F g^{-1} at 0.2 A g^{-1} was obtained by a ternary $MnFe_2O_4$/graphene/polyaniline nanocomposite fabricated by a facile two-step approach. The ternary nanocomposite also exhibited outstanding rate capability about 75.8% capacitance retention at 5 A g^{-1} and excellent cycling stability, 76.4% retention in capacitance after 5000 cycles. Specific capacitance of 307.2 F g^{-1} at 0.1 A g^{-1} has been achieved with symmetric supercapacitor. The device exhibited a maximum energy density of 13.5 W h kg^{-1}.

4. Zinc ferrite ($ZnFe_2O_4$)

Zn ferrite is widely used electrode material in supercapacitor because of its non-toxic nature, strong redox process, good chemical stability and high storage capacity of 2600 F g^{-1} [76–81]. Furthermore, the morphology of $ZnFe_2O_4$ can also be tuned such as nanoparticles [78, 82], nanorods [83] and nano-flakes [78, 80], offering large surface area for charge storage. M. M. Vadiyar et al. [84] have reported an empirical relationship between surface wettability and charge storing capacity for $ZnFe_2O_4$ nano-flake thin films. Different electrolytes, 1 M KOH, NaOH, LiOH and their combinations were chosen for this study. All the CV curves for $ZnFe_2O_4$ nano-flake recorded in the above electrolytes exhibited pseudocapacitive behavior in the scan range of 0.0 to −1.3 V. $ZnFe_2O_4$ nano-flakes exhibited larger area under the CV curves in 1 M KOH due to small hydrated K^+ and its fast intercalation and deintercalation on the electrode surface. This is found with good agreement with the smaller contact angle value of 12° and larger surface energy of 71 mJ m^{-2}.

$ZnFe_2O_4$ thin film synthesized by successive ionic layer adsorption and reaction (SILAR) method has shown good value of capacitance specific of 471 Fg^{-1} in aqueous electrolyte of 1 M NaOH at a scan rate of 5 mVs^{-1} [85]. The synthesized thin film of $ZnFe_2O_4$ was also used in solid-state symmetric supercapacitor which exhibited specific capacitance of 32 Fg^{-1} in voltage window of 1.0 V. A power density of 277 Wkg^{-1} with energy density of 4.47 $Whkg^{-1}$ was achieved with $ZnFe_2O_4$ thin film based supercapacitor. A specific capacitance of 615 F g^{-1} has been achieved for binder free $ZnFe_2O_4$ thin films at current density of 3 mA cm^{-2} [79]. The porous $ZnFe_2O_4$ thin film was tested in asymmetric supercapacitor as a negative electrode with positive electrode of Mn_3O_4. The device showed a specific capacitance of 81 F g^{-1} with energy and power density of 28 Wh kg^{-1} and 7.97 kW kg^{-1}, respectively. 74% retention was observed in capacitance after 3000 cycles.

$ZnFe_2O_4$ nanoparticles (size 20–30 nm) synthesized by combustion method was used for supercapacitor application [86]. The electrode showed a large maximum specific capacitance of 1235 F g^{-1} calculated at 1 mA cm^{-2}. The electrochemical performance of $ZnFe_2O_4$ material was also demonstrated in an asymmetric supercapacitor as negative electrode and nickel hydroxide as positive electrode. The device exhibited voltage window of 1.7 V with specific capacitance of 179 Fg^{-1} calculated at 2 mVs^{-1}. 3-D aligned $ZnFe_2O_4$ nano-flakes on flexible stainless steel mesh substrate have shown promising results as electrode in asymmetric supercapacitor with $Ni(OH)_2$ [87]. The asymmetric device exhibited large value of specific capacitance of 1625 F g^{-1} at 1 mA cm^{-2} with 97% retention in capacitance after 8000 cycles.

$ZnFe_2O_4$ microspheres synthesized by solvothermal approach demonstrated a specific capacitance of 131 Fg^{-1} [81]. The electrode of $ZnFe_2O_4$ microspheres could retain 92% capacitance after 1000 cycles. $ZnFe_2O_4$ anchored on multiwalled carbon nanotubes (CNT) yielded a high specific capacity of 217 mAh g^{-1} at 5 mV s^{-1} [88]. A solid-state symmetric supercapacitor with $ZnFe_2O_4$-CNT exhibited a highest specific energy of 12.80 Wh kg^{-1} and a specific power of 377.86 W kg^{-1}.

M. M. Vadiya et al. [78] have developed self-assembled $ZnFe_2O_4$ nanoflakes@ $ZnFe_2O_4$/C nanoparticles heterostructure electrode for high performance supercapacitor application. The hybrid electrode showed very high value of specific capacitance of 1884 Fg^{-1} determined at a current density of 5 mA cm^{-2}. A flexible asymmetric supercapacitor was also designed using $ZnFe_2O_4$ nano-flakes@ $ZnFe_2O_4$/C nanoparticles heterostructure as a negative electrode and reduced graphene oxide as a positive electrode. A specific capacitance of 347 F g^{-1} was achieved from the supercapacitor. The asymmetric supercapacitor exhibited an energy density of 81 Wh $kg^{-1}$and power density of 3.9 kW kg^{-1}. Only 2% loss in the capacitance was observed after 35000 cycles.

$ZnFe_2O_4$ nanoparticles were dispersed on nitrogen-doped reduced graphene for supercapacitor application [82]. The reduction of graphitic oxide, the doping of nitrogen to graphene and dispersion of $ZnFe_2O_4$ nanoparticles were achieved in a single process. The structure of $ZnFe_2O_4$/NRG exhibited a specific capacitance of 244 Fg^{-1} calculated at 0.5 Ag^{-1}. The electrode has also demonstrated good rate capability with retention of 131.5 Fg^{-1} capacitance at 10 Ag^{-1}. $ZnFe_2O_4$/NRG retained 83.8% capacitance after 5000 cycles. In this type of electrode, the graphene sheets provide high exposure of active sites for redox process and high dispersion of nanoparticles resulting good capacitive performance of the electrode [89, 90].

$ZnFe_2O_4$ nanorodes with rGO showed a specific capacitance of 1419 Fg^{-1} at scan rate of 10 mVs^{-1} in 2 M KOH solution. The electrode demonstrated good retention of capacitance about 93% after 5000 cycles. The improved electrochemical performance is due to the large surface area offered by rGO and good electrical conductivity [83]. The porous nano-flakes-$ZnFe_2O_4$ thin films demonstrated a larger capacitance of 768 Fg^{-1} at current density 5 mA cm^{-2} with energy density of 106 Wh kg^{-1} and power density of 18 kW kg^{-1} [76]. The electrode had good cycle stability about 88% retention of capacitance after 5000 cycles.

5. Nickel ferrite ($NiFe_2O_4$)

Among various metal ferrites, $NiFe_2O_4$ is one of the promising ferrite material for supercapacitor application. Several studies are available on $NiFe_2O_4$ and their nanocomposite as electrode in supercapacitor [46, 91–94]. $NiFe_2O_4$ particles with submicron-sized synthesized by the molten salt process demonstrated a specific capacitance of 18.5 F g^{-1} at a scan rate of 10 mV/s [95]. $NiFe_2O_4$ nanospheres showed a specific capacitance of 122 F g^{-1} at current density of 8.0 Ag^{-1} [26]. The results showed that the capacitance is increased with increasing the KOH concentration. $NiFe_2O_4$ nanospheres could store specific energy of 16.9 Wh kg^{-1} at a high current density of 8.0 Ag^{-1}.

Nagesh Kumar et al. [96] have synthesized mesoporous $NiFe_2O_4$ nanoparticles (size 10–15 nm) by one step hydrothermal method. The mesopores were distributed homogeneously on the surface of $NiFe_2O_4$ particle. A surface area of 148 m^2g^{-1} was calculated for the mesoporous $NiFe_2O_4$ nanoparticles. It exhibited high value of specific capacitance of 1040 Fg^{-1} at 1 Ag^{-1} in a three-electrode configuration with 2 M KOH electrolyte. However, 30% loss in capacitance was observed for $NiFe_2O_4$ nanoparticles after 500 cycles. $NiFe_2O_4$ synthesized by combustion route showed a specific capacitance of 454 Fg^{-1} with good cycle stability for 1000 charging-discharging cycles [46].

To improve the capacitive properties, graphene based materials have been mixed with $NiFe_2O_4$ [97–99]. Soam et al. [98] obtained a specific capacitance of 207 Fg^{-1} from ferrite/graphene nanocomposite in 1 M Na_2SO_4 electrolyte (**Figure 3**). This value of capacitance was observed about 4 times greater than $NiFe_2O_4$ electrode. $NiFe_2O_4$ with graphene nanosheets exhibited a stable capacitance about 95% over 1000 cycles. Numerous pores in the electrode might be responsible for improved

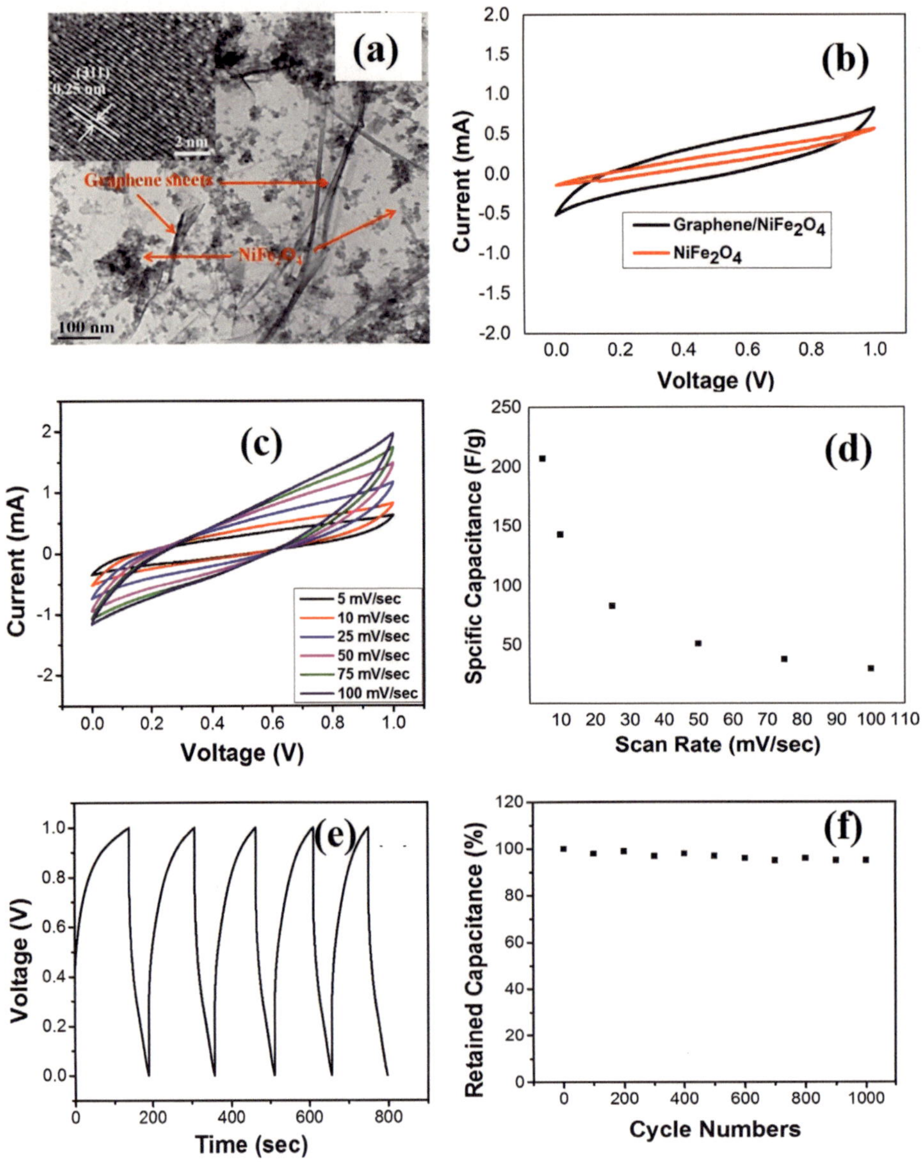

Figure 3.
(a) TEM images of graphene/NiFe$_2$O$_4$ nanocomposite. NiFe$_2$O$_4$ nanoparticles have good contact with graphene, providing fast charge transportation within the electrode (b) CV curves of graphene and NiFe$_2$O$_4$ recorded at 5 mVs^{-1}. The nanocomposite of graphene/NiFe$_2$O$_4$ exhibited larger area under CV curve, indicating better charge storage capacity than NiFe$_2$O$_4$. (c) CV curves at different scan rates for graphene/NiFe$_2$O$_4$ nanocomposite. (d) Capacitance versus scan rate for the graphene/NiFe$_2$O$_4$ nanocomposite, the electrode exhibited specific capacitance in the range of 207-30 Fg^{-1} at scan rates of 5-100 mVs^{-1} and (e) charging/discharging curves with constant current of 0.5 mA. (f) Cycle stability test performed over 1000 cycles [98].

electrochemical performance of NiFe$_2$O$_4$. Zhuo Wang et al. [100] have studied the reduce graphene oxide–NiFe$_2$O$_4$ nanocomposites for supercapacitor application. rGO-NiFe$_2$O$_4$ nanocomposites were prepared by hydrothermal process with varying the pH value of solution (8, 10, 12 and 14). rGO-NiFe$_2$O$_4$ synthesized with pH value of 10 exhibited the largest surface area of 459.6 m^2 g^{-1}. A specific capacitance of 218.47 Fg^{-1} was achieved for the rGO-NiFe$_2$O$_4$ (pH -10) at 5 mV/s^{-1}, which is the largest among all the samples.

Addition of conducting network of PANI to $NiFe_2O_4$ improved the electrochemical performance of PANI-$NiFe_2O_4$ nanocomposite electrode [101]. A specific capacitance of 448 Fg^{-1} was achieved with PANI-$NiFe_2O_4$. The electrode showed 80% retention in the capacitance after 1000 cycles at the rate of 10 $mAcm^{-2}$. A composite of mesoporous $NiFe_2O_4$ with multiwall carbon nanotubes (MWCNTs) prepared via hexamethylene tetramine (HMT) assisted one pot hydrothermal process exhibited large value of specific capacitance, 1291 F g^{-1} determined at 1 A g^{-1} [92]. The electrode showed capacitance retention of 81% over 500 charge–discharge cycles in 2 M KOH electrolyte. The asymmetric device with $NiFe_2O_4$/CNT nanocomposite as cathode and N-doped graphene as anode demonstrated a specific capacitance of 66 F g^{-1} with energy density of 23 W h kg^{-1} and power density of 872 W kg^{-1}.

$NiFe_2O_4$ nanoparticles grown on a flexible carbon cloth substrate via hydrothermal method demonstrated a high capacitance 1135.5 F g^{-1} in 1 M H_2SO_4 electrolyte and 922.6 F g^{-1} in 6 M KOH electrolyte with current density of 2 $mAcm^{-2}$ [93]. The large capacitance can be attributed to the conductive 3D network of carbon cloth and large surface area for $NiFe_2O_4$ nanoparticles. The binder free electrode of $NiFe_2O_4$ nanocone forest on carbon textile (NFO-CT) exhibited specific capacitance of 697 F g^{-1} calculated by CV at scan rate of 5 mV s^{-1} [94]. Further, a solid state supercapacitor of NFO-CT also demonstrated good value of capacitance of 584 F g^{-1} at 5 mV s^{-1}. Moreover, the device showed good cycle stability with 93.57% capacitance retention over 10,000 cycles. These results indicate that NFO-CT may be a promising candidate for high performance supercapacitor. The capacitance of $NiFe_2O_4$ was also observed to be dependent on the synthesis process [91]. The $NiFe_2O_4$ synthesized by combustion, polyol-mediated and sol–gel methods have different morphology and consequently different value of capacitance. A high specific capacitance value of 97.5 Fg^{-1} was obtained from sol–gel synthesized method. The size of grains and pores are smaller for sol–gel synthesized $NiFe_2O_4$ which could be the reason for better value of capacitance.

1D $NiFe_2O_4$/graphene composites prepared via hydrothermal process exhibited specific capacitance of 481.3 F g^{-1} at a current density of 0.1 A g^{-1} [97]. The 1D $NiFe_2O_4$/graphene electrode maintained 298.2 F g^{-1} capacitance upon increasing the current density to 10 A g^{-1}. The electrode demonstrated outstanding cycle stability over 10000 cycles (about 1% degradation in capacitance). On the other hand, 40% loss of capacitance was observed for $NiFe_2O_4$ electrode (125 to 75 F g^{-1}). The excellent electrochemical performance of $NiFe_2O_4$/graphene composites electrode is due to the conducting network of graphene and large number of redox active site from $NiFe_2O_4$. Ternary nitrogen-doped graphene/nickel ferrite/polyaniline (NGNP) nanocomposite showed specific capacitance of 645.0 F g^{-1} at 1 mV s^{-1} [99]. In a two-electrode symmetric system, the energy density and power density were determined to be 92.7 W h kg^{-1} and 110.8 W kg^{-1}, respectively. About 90% retention in capacitance was seen after 10,000 cycles. The electrochemical behavior of NGNP is improved due to combined effects of EDLC and pseudocapacitor.

6. Conclusions

In this chapter, electrochemical performance of selected ferrites ($CoFe_2O_4$, $MnFe_2O_4$, $ZnFe_2O_4$, and $NiFe_2O_4$) and their nanocomposites with conducting carbon network for supercapacitor has been reviewed. Their synthesis process was also highlighted. The surface morphology of these materials plays an important role in supercapacitor. These materials store the charge by redox process. However, poor electrical conductivity is the main limitation to be used them in fast charging/

discharging supercapacitor. In this regard, their nanocomposites with graphene enhanced the electrochemical performance. Nano-flakes type structure with graphene exhibited great electrochemical performance in supercapacitor.

Conflict of interest

The authors declare no conflict of interest.

Author details

Ankur Soam
Institute of Technical Education and Research, Siksha 'O' Anusandhan (Deemed to be University), Bhubaneswar, India

*Address all correspondence to: ankursoam007@gmail.com

References

[1] T. Kim, W. Song, D. Y. Son, L. K. Ono, and Y. Qi, "Lithium-ion batteries: outlook on present, future, and hybridized technologies," J. Mater. Chem. A, vol. 7, no. 7, pp. 2942-2964, 2019, doi: 10.1039/C8TA10513H.

[2] N. Nitta, F. Wu, J. T. Lee, and G. Yushin, "Li-ion battery materials: Present and future," Mater. Today, vol. 18, no. 5, pp. 252-264, 2015, doi: 10.1016/j.mattod.2014.10.040.

[3] H. J. Kim et al., "A comprehensive review of li-ion battery materials and their recycling techniques", vol. 9, no. 7, 1161, 2020, doi: org/10.3390/electronics9071161

[4] H. Ning, J. H. Pikul, R. Zhang, X. Li, S. Xu, J. Wang, J. A. Rogers, William P. King, and Paul V. Braun, "Holographic patterning of high-performance on-chip 3D lithium-ion microbatteries" Proceedings of the National Academy of Sciences, vol. 112, no. 21, 6573-6578, 2015, doi: 10.1073/pnas.1423889112

[5] B. Pu et al., "Flexible supercapacitors based on carbon nanomaterials," J. Mater. Chem. A Mater. energy Sustain., vol. 3, no. 1, pp. 22507-22541, 2014, doi: 10.1016/j.mattod.2015.01.002.

[6] M. Huang, F. Li, F. Dong, Y. X. Zhang, and L. L. Zhang, "MnO2-based nanostructures for high-performance supercapacitors," J. Mater. Chem. A, vol. 3, no. 43, pp. 21380-21423, 2015, doi: 10.1039/c5ta05523g.

[7] M. Zhi, C. Xiang, J. Li, M. Li, and N. Wu, "Nanostructured carbon – metal oxide composite electrodes for supercapacitors : a review," pp. 72-88, 2013, doi: 10.1039/c2nr32040a.

[8] A. Soam, P. Kavle, A. Kumbhar, and R. O. Dusane, "Performance enhancement of micro-supercapacitor by coating of graphene on silicon nanowires at room temperature," Curr. Appl. Phys., vol. 17, no. 2, 2017, doi: 10.1016/j.cap.2016.11.011.

[9] R. Kumar, A. Soam, R. Hussain, I. Mansuri, and V. Sahajwalla, "Carbon coated iron oxide (CC-IO) as high performance electrode material for supercapacitor applications," J. Energy Storage, vol. 32, no. August, p. 101737, 2020, doi: 10.1016/j.est.2020.101737.

[10] M. Beidaghi and Y. Gogotsi, "Capacitive energy storage in micro-scale devices: Recent advances in design and fabrication of micro-supercapacitors," Energy Environ. Sci., vol. 7, no. 3, pp. 867-884, 2014, doi: 10.1039/c3ee43526a.

[11] A. Ben Amar, A. B. Kouki, and H. Cao, "Power Approaches for Implantable Medical Devices," Sensors (Basel), vol. 15, no. 11, pp. 28889-28914, 2015, doi: 10.3390/s151128889.

[12] X. Wei et al., "Flexible fiber energy storage and integrated devices : recent progress and perspectives," J. Mater. Chem. A Mater. energy Sustain., vol. 3, no. 1, pp. 1553-1565, 2014, doi: 10.1016/j.bios.2009.12.001.Emerging.

[13] M. I. A. A. Maksoud et al., "Advanced materials and technologies for supercapacitors used in energy conversion and storage : a review", Environmental Chemistry Letters, vol. 19, no. 1, pp. 375-439, 2021, doi: org/10.1007/s10311-020-01075-w.

[14] Y. H. Joung, "Development of Implantable Medical Devices : Int Neurourol J. 2013 Sep;17(3):98-106. doi: 10.5213/inj.2013.17.3.98.

[15] W. Zheng, "A review for aqueous electrochemical supercapacitors," vol. 3, no. May, pp. 1-11, 2015, doi: 10.3389/fenrg.2015.00023.

[16] P. Sundriyal and S. Bhattacharya, "Energy Harvesting Techniques for Powering Wireless Sensor Networks in Aircraft Applications: A Review. In: Bhattacharya S., Agarwal A., Prakash O., Singh S. (eds) Sensors for Automotive and Aerospace Applications. Energy, Environment, and Sustainability. Springer, Singapore, 2019, doi: 10.1007/978-981-13-3290-6.

[17] H. Gao, S. Liu, Y. Li, E. Conte, and Y. Cao, "A critical review of spinel structured iron cobalt oxides based materials for electrochemical energy storage and conversion," Energies, vol. 10, no. 11, 2017, doi: 10.3390/en10111787.

[18] A. Yan, X. Wang, and J. Cheng, "Research Progress of NiMn Layered Double Hydroxides for Supercapacitors : A Review," Nanomaterials (Basel), vol. 8, no. 10, pp. 747, 2018, doi: 10.3390/nano8100747.

[19] A. A. El-Moneim, "Two steps synthesis approach of MnO2/graphene nanoplates/graphite composite electrode for supercapacitor application," Mater. Today Energy, vol. 3, pp. 24-31, 2017, doi: 10.1016/j.mtener.2017.02.004.

[20] Y. Hou, Y. Cheng, T. Hobson, and J. Liu, "Design and Synthesis of Hierarchical MnO2 Nanospheres/Carbon Nanotubes/Conducting Polymer Ternary Composite for High Performance Electrochemical Electrodes," Nano Lett, vol. 10, no. 7, pp. 2727-2733, 2010, doi: 10.1021/nl101723g.

[21] A. D. Smith et al., "Carbon-Based Electrode Materials for Microsupercapacitors in Self-Powering Sensor Networks : Present and Future Development." Sensors, vol. 19, no.19, pp. 4231, 2019, doi: org/10.3390/s19194231.

[22] A. Soam, N. Arya, A. Singh, and R. Dusane, "Fabrication of silicon nanowires based on-chip micro-supercapacitor," Chem. Phys. Lett., vol. 678, pp. 46-50, 2017, doi: 10.1016/j.cplett.2017.04.019.

[23] A. Soam, K. Parida, R. Kumar, P. kavle, and R. O. Dusane, "Silicon-MnO<inf>2</inf> core-shell nanowires as electrodes for micro-supercapacitor application," Ceram. Int., vol. 45, no. 15, 2019, doi: 10.1016/j.ceramint.2019.06.127.

[24] R. Kumar, A. Soam, R. O. Dusane, and P. Bhargava, "Sucrose derived carbon coated silicon nanowires for supercapacitor application," J. Mater. Sci. Mater. Electron., vol. 29, pp. 1947-1954, 2018, doi: 10.1007/s10854-017-8105-x.

[25] A. Sarkar, A. K. Singh, D. Sarkar, G. G. Khan, and K. Mandal, "Three-Dimensional Nanoarchitecture of BiFeO 3 Anchored TiO 2 Nanotube Arrays for Electrochemical Energy Storage and Solar Energy Conversion," ACS Sustainable Chem. Eng., vol. 3, no. 9, pp. 2254-2263, 2015, doi: 10.1021/acssuschemeng.5b00519.

[26] A. Ghasemi, M. Kheirmand, and H. Heli, "Synthesis of Novel NiFe2O4 Nanospheres for High Performance Pseudocapacitor Applications," Russian Journal of Electrochemistry, vol. 55, no. 3, pp. 341-349, 2019, doi: 10.1134/S1023193519020022.

[27] S. L. Candelaria et al., "Nanostructured carbon for energy storage and conversion," Nano Energy, vol. 1, no. 2, pp. 195-220, 2012, doi: 10.1016/j.nanoen.2011.11.006.

[28] R. Kumar, A. Soam, and V. Sahajwalla, " Carbon coated cobalt oxide (CC-CO 3 O 4) as electrode material for supercapacitor applications ," Mater. Adv., vol. 2, no. 9, pp. 2918-2923, 2021, doi: 10.1039/d1ma00120e.

[29] R. Kumar, B. K. Singh, A. Soam, S. Parida, and P. Bhargava, "In-situ carbon coated manganese oxide nanorods

(ISCC-MnO <inf>2</inf> NRs) as an electrode material for supercapacitors," Diam. Relat. Mater., vol. 94, pp. 110-117, 2019, doi: 10.1016/j. diamond.2019.03.003.

[30] X. Wang, D. Wu, X. Song, W. Du, X. Zhao, and D. Zhang, "Review on Carbon / Polyaniline Hybrids : Design and Synthesis for Supercapacitor," Molecules, vol. 24, no. 12, pp. 2263, 2019, doi: 10.3390/molecules24122263 2019.

[31] H. Pan, J. Li, and Y. Ping, "Carbon Nanotubes for Supercapacitor," Nanoscale Research Letters, vol. 5, pp. 654-668, 2010, doi: 10.1007/s11671-009-9508-2.

[32] V. C. Bose and V. Biju, "Mixed valence nanostructured Mn<inf>3</inf>O<inf>4</inf> for supercapacitor applications," Bull. Mater. Sci., vol. 38, no. 4, pp. 865-873, 2015, doi: 10.1007/s12034-015-0906-z.

[33] Z. S. Wu, W. Ren, D. W. Wang, F. Li, B. Liu and H. M. Cheng, "High-Energy MnO 2 Nanowire/Graphene and Graphene Asymmetric Electrochemical Capacitors," ACS Nano, vol. 4, no. 10, pp. 5835-5842, 2010, doi: org/10.1021/nn101754k

[34] S. Nayak et al., "Sol–gel synthesized BiFeO3–Graphene nanocomposite as efficient electrode for supercapacitor application," J. Mater. Sci. Mater. Electron., vol. 29, no. 11, pp. 9361-9368, 2018, doi: 10.1007/s10854-018-8967-6.

[35] M. Notarianni, J. Liu, F. Mirri, M. Pasquali, and N. Motta, "Graphene-based supercapacitor with carbon nanotube film as highly efficient current collector," Nanotechnology, vol. 25, no. 43, 2014, doi: 10.1088/0957-4484/25/43/435405.

[36] J. J. Yoo et al., "Ultrathin Planar Graphene Supercapacitors," Nano Lett., vol. 11, no. 4, pp. 1423-1427, 2011, doi: org/10.1021/nl200225j.

[37] E. Frackowiak, K. Metenier, V. Bertagna, and F. Beguin, "Supercapacitor electrodes from multiwalled carbon nanotubes," Appl. Phys. Lett., vol. 77, no. 15, pp. 2421-2423, 2010, doi: org/10.1063/1.1290146.

[38] Y. Wang et al., "Supercapacitor Devices Based on Graphene Materials," J. Phys. Chem. C, vol. 113, no. 30, pp. 13103-13107, 2009, doi.org/10.1021/jp902214f.

[39] M. I. A. Abdel Maksoud et al., Advanced materials and technologies for supercapacitors used in energy conversion and storage: a review, Environ Chem Lett, vol. 19, pp. 375-439, 2021, doi.org/10.1007/s10311-020-01075-w.

[40] M. Zhi, C. Xiang, J. Li, M. Li, and N. Wu, "Nanostructured carbon-metal oxide composite electrodes for supercapacitors: A review," Nanoscale, vol. 5, no. 1, pp. 72-88, 2013, doi: 10.1039/c2nr32040a.

[41] Z. Lin et al., "Materials for supercapacitors: When Li-ion battery power is not enough", Materials Today, vol. 21, pp. 419-436, 2018, doi: org/10.1016/j.mattod.2018.01.035.

[42] N. Choudhary et al., "Asymmetric Supercapacitor Electrodes and Devices," Adv. Mater., vol. 29, no. 21, pp. 1605336, 2017, doi: 10.1002/adma.201605336.

[43] A. Soam and R. Kumar, "Preparation of MnO2 Nanoparticles by a solution based approach for electrochemical capacitor," Surf. Rev. Lett., vol. 27, no. 8, pp. 21-26, 2020, doi: 10.1142/S0218625X19501993.

[44] Z. S. Iro, C. Subramani, and S. S. Dash, "A brief review on electrode materials for supercapacitor," Int. J. Electrochem. Sci., vol. 11, no. 12, pp.

10628-10643, 2016, doi: 10.20964/2016.12.50.

[45] X. Yao, J. Kong, D. Zhou, C. Zhao, R. Zhou, and X. Lu, "Mesoporous zinc ferrite / graphene composites : Towards ultra-fast and stable anode for lithium-ion batteries," Carbon, vol. 79, pp. 493-499, 2014, doi: 10.1016/j.carbon.2014.08.007.

[46] V. Venkatachalam and R. Jayavel, "Novel Synthesis of Ni-Ferrite (NiFe2O4) Electrode Material for Supercapacitor Applications," AIP Conference Proceedings, vol. 1665, no. 140016, 2015, doi: 10.1063/1.4918225.

[47] P. V. Shinde, N. M. Shinde, R. S. Mane, and K. H. Kim, Chapter 5 - Ferrites for Electrochemical Supercapacitors, Editor(s): Rajaram S. Mane, Vijaykumar V. Jadhav, In Micro and Nano Technologies, Spinel Ferrite Nanostructures for Energy Storage Devices, Elsevier, 2020, pages 83-122, ISBN 9780128192375, doi: org/10.1016/B978-0-12-819237-5.00005-5.

[48] A. Soam, R. Kumar, D. Thatoi, and M. Singh, "Electrochemical Performance and Working Voltage Optimization of Nickel Ferrite/Graphene Composite based Supercapacitor," J. Inorg. Organomet. Polym. Mater., vol. 30, no. 9, pp. 3325-3331, 2020, doi: 10.1007/s10904-020-01540-7.

[49] S. Kuo and N. Wu, "Study on ferrites for supercapacitor application," 56th Annu. Meet. Int. Soc. Electrochem., no. December, pp. 1-5, 2015.

[50] S. F. Shaikh, M. Ubaidullah, R. S. Mane, and A. M. Al-Enizi, Chapter 4 - Types, Synthesis methods and applications of ferrites, Editor(s): Rajaram S. Mane, Vijaykumar V. Jadhav, In Micro and Nano Technologies, Spinel Ferrite Nanostructures for Energy Storage Devices, Elsevier, 2020, Pages 51-82, ISBN 9780128192375, doi. org/10.1016/B978-0-12-819237-5.00004-3.

[51] V. V Jadhav, M. K. Zate, S. Liu, and M. Naushad, "Mixed-phase bismuth ferrite nanoflake electrodes for supercapacitor application," Appl. Nanosci., vol. 6, no. 4, pp. 511-519, 2016, doi: 10.1007/s13204-015-0469-8.

[52] V. S. Kumbhar, A. D. Jagadale, N. M. Shinde, and C. D. Lokhande, "Chemical synthesis of spinel cobalt ferrite (CoFe 2 O 4) nano-flakes for supercapacitor application," Appl. Surf. Sci., vol. 259, pp. 39-43, 2012, doi: 10.1016/j.apsusc.2012.06.034.

[53] B. Wang et al., "Electrocapacitive properties of MnFe2O4 electrodes in aqueous LiNO3 electrolyte with surfactants," Int. J. Electrochem. Sci., vol. 8, no. 7, pp. 8966-8977, 2013.

[54] N. Masunga, O. K. Mmelesi, K. K. Kefeni, and B. B. Mamba, "Recent advances in copper ferrite nanoparticles and nanocomposites synthesis, magnetic properties and application in water treatment: Review," J. Environ. Chem. Eng., vol. 7, no. 3, p. 103179, 2019, doi: 10.1016/j.jece.2019.103179.

[55] A. Shanmugavani, D. Kalpana, and R. K. Selvan, "Electrochemical properties of CoFe2O4 nanoparticles as negative and Co(OH)2 and Co2Fe(CN)6 as positive electrodes for supercapacitors," Mater. Res. Bull., vol. 71, pp. 133-141, 2015, doi: 10.1016/j.materresbull.2015.04.018.

[56] C. Paper, S. Patil, M. Chithra, S. C. Sahoo, and P. B. Patil, "Cobalt Ferrite Nanoparticles for Supercapacitor, AIP Conference Proceedings vol. 2265, no. 030162, 2020, https://doi.org/10.1063/5.0017184.

[57] A. E. Elkholy, F. El-Taib Heakal, and N. K. Allam, "Nanostructured spinel manganese cobalt ferrite for high-performance supercapacitors," RSC

Adv., vol. 7, no. 82, pp. 51888-51895, 2017, doi: 10.1039/c7ra11020k.

[58] S. G. Kandalkar, J. L. Gunjakar, and C. D. Lokhande, "Preparation of cobalt oxide thin films and its use in supercapacitor application," Applied Surface Science, vol. 254, pp. 5540-5544, 2008, doi: 10.1016/j.apsusc.2008.02.163.

[59] S. M. Alshehri, J. Ahmed, A. N. Alhabarah, T. Ahamad, and T. Ahmad, "Nitrogen-Doped Cobalt Ferrite/Carbon Nanocomposites for Supercapacitor Applications," ChemElectroChem, vol. 4, no. 11, pp. 2952-2958, 2017, doi: 10.1002/celc.201700602.

[60] S. J. Pawar, S. M. Patil, M. Chithra, S. C. Sahoo, and P. B. Patil, "Cobalt ferrite nanoparticles for supercapacitor application," AIP Conf. Proc., vol. 2265, no. 030162, 2020, doi: 10.1063/5.0017184.

[61] Y. X. Zhang, X. D. Hao, Z. P. Diao, J. Li, and Y. M. Guan, "One-pot controllable synthesis of flower-like CoFe2O 4/FeOOH nanocomposites for high-performance supercapacitors," Mater. Lett., vol. 123, pp. 229-234, 2014, doi: 10.1016/j.matlet.2014.02.103.

[62] V. A. Jundale, D. A. Patil, G. Y. Chorage, and A. A. Yadav, "Mesoporous cobalt ferrite thin film for supercapacitor applications," Mater. Today Proc., vol. 43, pp. 2711-2715, 2020, doi: 10.1016/j.matpr.2020.06.204.

[63] H. Kennaz et al., "Synthesis and electrochemical investigation of spinel cobalt ferrite magnetic nanoparticles for supercapacitor application," J. Solid State Electrochem., vol. 22, no. 3, pp. 835-847, 2018, doi: 10.1007/s10008-017-3813-y.

[64] P. He, K. Yang, W. Wang, F. Dong, L. Du, and Y. Deng, "Reduced graphene oxide-CoFe2O4 composites for supercapacitor electrode," Russ. J. Electrochem., vol. 49, no. 4, pp. 359-364, 2013, doi: 10.1134/S1023193513040101.

[65] P. Xiong, H. Huang, and X. Wang, "Design and synthesis of ternary cobalt ferrite/graphene/polyaniline hierarchical nanocomposites for high-performance supercapacitors" J. Power Sources, vol. 245, pp. 937-946, 2014, doi: 10.1016/j.jpowsour.2013.07.064.

[66] W. Cai, T. Lai, W. Dai, and J. Ye, "A facile approach to fabricate flexible all-solid-state supercapacitors based on MnFe2O4/graphene hybrids," J. Power Sources, vol. 255, pp. 170-178, 2014, doi: 10.1016/j.jpowsour.2014.01.027.

[67] B. Bashir et al., "Copper doped manganese ferrites nanoparticles anchored on graphene nano-sheets for high performance energy storage applications," J. Alloys Compd., vol. 695, pp. 881-887, 2017, doi: 10.1016/j.jallcom.2016.10.183.

[68] L. H. Nonaka, T. S. D. Almeida, C. B. Aquino, S. H. Domingues, R. V. Salvatierra, and V. H. R. Souza, "Crumpled Graphene Decorated with Manganese Ferrite Nanoparticles for Hydrogen Peroxide Sensing and Electrochemical Supercapacitors," ACS Appl. Nano Mater., vol. 3, no. 5, pp. 4859-4869, 2020, doi: 10.1021/acsanm.0c01012.

[69] V. Vignesh, K. Subramani, M. Sathish, and R. Navamathavan, "Electrochemical investigation of manganese ferrites prepared via a facile synthesis route for supercapacitor applications," Colloids Surfaces A Physicochem. Eng. Asp., vol. 538, pp. 668-677, 2018, doi: 10.1016/j.colsurfa.2017.11.045.

[70] P. Guo et al., "Electrochemical properties of colloidal nanocrystal assemblies of manganese ferrite as the electrode materials for supercapacitors," J. Mater. Sci., vol. 52, no. 9, pp.

5359-5365, 2017, doi: 10.1007/s10853-017-0778-2.

[71] S.-L. Kuo, J.-F. Lee, and N.-L. Wu, "Study on Pseudocapacitance Mechanism of Aqueous MnFe[sub 2]O[sub 4] Supercapacitor," J. Electrochem. Soc., vol. 154, no. 1, p. A34, 2007, doi: 10.1149/1.2388743.

[72] R. Wang et al., "Electrochemical properties of manganese ferrite-based supercapacitors in aqueous electrolyte: The effect of ionic radius," Colloids Surfaces A Physicochem. Eng. Asp., vol. 457, no. 1, pp. 94-99, 2014, doi: 10.1016/j.colsurfa.2014.05.059.

[73] I. Kotutha, E. Swatsitang, W. Meewassana, and S. Maensiri, "One-pot hydrothermal synthesis, characterization, and electrochemical properties of rGO/MnFe2O4 nanocomposites," Jpn. J. Appl. Phys., vol. 54, no. 6, 2015, doi: 10.7567/JJAP.54.06FH10.

[74] P. Xiong, C. Hu, Y. Fan, W. Zhang, J. Zhu, and X. Wang, "Ternary manganese ferrite/graphene/polyaniline nanostructure with enhanced electrochemical capacitance performance," J. Power Sources, vol. 266, pp. 384-392, 2014, doi: 10.1016/j.jpowsour.2014.05.048.

[75] K. V. Sankar and R. K. Selvan, "The ternary MnFe2O4/graphene/polyaniline hybrid composite as negative electrode for supercapacitors," J. Power Sources, vol. 275, pp. 399-407, 2015, doi: 10.1016/j.jpowsour.2014.10.183.

[76] M. M. Vadiyar et al., "Mechanochemical growth of a porous ZnFe2O4 nano-flake thin film as an electrode for supercapacitor application," RSC Adv., vol. 5, no. 57, pp. 45935-45942, 2015, doi: 10.1039/c5ra07588b.

[77] M. M. Vadiyar, S. S. Kolekar, J. Y. Chang, A. A. Kashale, and A. V. Ghule, "Reflux Condensation Mediated Deposition of Co3O4 Nanosheets and ZnFe2O4 Nanoflakes Electrodes for Flexible Asymmetric Supercapacitor," Electrochim. Acta, vol. 222, pp. 1604-1615, 2016, doi: 10.1016/j.electacta.2016.11.146.

[78] M. M. Vadiyar, S. S. Kolekar, J. Y. Chang, Z. Ye, and A. V. Ghule, "Anchoring Ultrafine ZnFe2O4/C Nanoparticles on 3D ZnFe2O4 Nanoflakes for Boosting Cycle Stability and Energy Density of Flexible Asymmetric Supercapacitor," ACS Appl. Mater. Interfaces, vol. 9, no. 31, pp. 26016-26028, 2017, doi: 10.1021/acsami.7b06847.

[79] M. M. Vadiyar, S. S. Kolekar, N. G. Deshpande, J. Y. Chang, A. A. Kashale, and A. V. Ghule, "Binder-free chemical synthesis of ZnFe2O4 thin films for asymmetric supercapacitor with improved performance," Ionics (Kiel)., vol. 23, no. 3, pp. 741-749, 2017, doi: 10.1007/s11581-016-1833-8.

[80] M. M. Vadiyar, S. C. Bhise, S. K. Patil, S. S. Kolekar, J. Y. Chang, and A. V. Ghule, "Comparative Study of Individual and Mixed Aqueous Electrolytes with ZnFe2O4 Nano–flakes Thin Film as an Electrode for Supercapacitor Application," ChemistrySelect, vol. 1, no. 5, pp. 959-966, 2016, doi: 10.1002/slct.201600151.

[81] M. Zhu, X. Zhang, Y. Zhou, C. Zhuo, J. Huang, and S. Li, "Facile solvothermal synthesis of porous ZnFe2O4 microspheres for capacitive pseudocapacitors," RSC Adv., vol. 5, no. 49, pp. 39270-39277, 2015, doi: 10.1039/c5ra00447k.

[82] L. Li et al., "Uniformly Dispersed ZnFe2O4 Nanoparticles on Nitrogen-Modified Graphene for High-Performance Supercapacitor as Electrode," Scientific Reports, vol. 7, no. 43116, 2017, doi: 10.1038/srep43116.

[83] M. B. Askari, P. Salarizadeh, M. Seifi, M. H. Ramezan zadeh, and A. Di Bartolomeo, "ZnFe2O4 nanorods on reduced graphene oxide as advanced supercapacitor electrodes," J. Alloys Compd., vol. 860, p. 158497, 2021, doi: 10.1016/j.jallcom.2020.158497.

[84] M. M. Vadiyar et al., "Contact angle measurements: A preliminary diagnostic tool for evaluating the performance of ZnFe2O4 nano-flake based supercapacitors," Chem. Commun., vol. 52, no. 12, pp. 2557-2560, 2016, doi: 10.1039/c5cc08373g.

[85] S. S. Raut and B. R. Sankapal, "First report on synthesis of ZnFe2O4 thin film using successive ionic layer adsorption and reaction: Approach towards solid-state symmetric supercapacitor device," Electrochim. Acta, vol. 198, pp. 203-211, 2016, doi: 10.1016/j.electacta.2016.03.059.

[86] A. Shanmugavani and R. K. Selvan, "Synthesis of ZnFe2O4 nanoparticles and their asymmetric configuration with Ni(OH)2 for a pseudocapacitor," RSC Adv., vol. 4, no. 51, pp. 27022-27029, 2014, doi: 10.1039/c4ra01793e.

[87] M. M. Vadiyar, S. C. Bhise, S. S. Kolekar, J. Y. Chang, K. S. Ghule, and A. V. Ghule, "Low cost flexible 3-D aligned and cross-linked efficient ZnFe2O4 nano-flakes electrode on stainless steel mesh for asymmetric supercapacitors," J. Mater. Chem. A, vol. 4, no. 9, pp. 3504-3512, 2016, doi: 10.1039/c5ta09022a.

[88] S. S. Raut, B. R. Sankapal, M. S. A. Hossain, S. Pradhan, R. R. Salunkhe, and Y. Yamauchi, "Zinc Ferrite Anchored Multiwalled Carbon Nanotubes for High-Performance Supercapacitor Applications," Eur. J. Inorg. Chem., vol. 2018, no. 2, pp. 137-142, 2018, doi: 10.1002/ejic.201700836.

[89] A. Soam, C. Mahender, R. Kumar, and M. Singh, "Power performance of BFO-graphene composite electrodes based supercapacitor," Mater. Res. Express, vol. 6, no. 2, 2019, doi: 10.1088/2053-1591/aaf125.

[90] L. Mao, K. Zhang, H. S. On Chan, and J. Wu, "Nanostructured MnO2/graphene composites for supercapacitor electrodes: The effect of morphology, crystallinity and composition," J. Mater. Chem., vol. 22, no. 5, pp. 1845-1851, 2012, doi: 10.1039/c1jm14503g.

[91] S. Anwar, K. S. Muthu, V. Ganesh, and N. Lakshminarasimhan, "A Comparative Study of Electrochemical Capacitive Behavior of NiFe2O4 Synthesized by Different Routes," J. Electrochem. Soc., vol. 158, no. 8, p. A976, 2011, doi: 10.1149/1.3601863.

[92] N. Kumar, A. Kumar, G. M. Huang, W. W. Wu, and T. Y. Tseng, "Facile synthesis of mesoporous NiFe2O4/CNTs nanocomposite cathode material for high performance asymmetric pseudocapacitors," Appl. Surf. Sci., vol. 433, pp. 1100-1112, 2018, doi: 10.1016/j.apsusc.2017.10.095.

[93] Z. Y. Yu, L. F. Chen, and S. H. Yu, "Growth of NiFe2O4 nanoparticles on carbon cloth for high performance flexible supercapacitors," J. Mater. Chem. A, vol. 2, no. 28, pp. 10889-10894, 2014, doi: 10.1039/c4ta00492b.

[94] M. S. Javed, C. Zhang, L. Chen, Y. Xi, and C. Hu, "Hierarchical mesoporous NiFe2O4 nanocone forest directly growing on carbon textile for high performance flexible supercapacitors," J. Mater. Chem. A, vol. 4, no. 22, pp. 8851-8859, 2016, doi: 10.1039/c6ta01893a.

[95] B. Senthilkumar, R. Kalai Selvan, P. Vinothbabu, I. Perelshtein, and A. Gedanken, "Structural, magnetic, electrical and electrochemical properties of NiFe2O4 synthesized by the molten salt technique," Mater. Chem. Phys., vol. 130, no. 1-2, pp.

285-292, 2011, doi: 10.1016/j.matchemphys.2011.06.043.

[96] N. Kumar, A. Kumar, S. Chandrasekaran, and T. Y. Tseng, "Synthesis of Mesoporous NiFe2O4 Nanoparticles for Enhanced Supercapacitive Performance," J. Clean Energy Technol., vol. 6, no. 1, pp. 51-55, 2018, doi: 10.18178/jocet.2018.6.1.435.

[97] M. Fu, W. Chen, X. Zhu, and Q. Liu, "One-step preparation of one dimensional nickel ferrites / graphene composites for supercapacitor electrode with excellent cycling stability," J. Power Sources, vol. 396, no. March, pp. 41-48, 2018, doi: 10.1016/j.jpowsour.2018.06.019.

[98] A. Soam et al., "Synthesis of Nickel Ferrite Nanoparticles Supported on Graphene Nanosheets as Composite Electrodes for High Performance Supercapacitor," ChemistrySelect, vol. 4, no. 34, pp. 9952-9958, 2019, doi: 10.1002/slct.201901117.

[99] W. Wang, Q. Hao, W. Lei, X. Xia, and X. Wang, "Ternary nitrogen-doped graphene / nickel ferrite / polyaniline nanocomposites for high-performance supercapacitors," J. Power Sources, vol. 269, pp. 250-259, 2014, doi: 10.1016/j.jpowsour.2014.07.010.

[100] Z. Wang, X. Zhang, Y. Li, Z. Liu, and Z. Hao, "Synthesis of graphene-NiFe2O4 nanocomposites and their electrochemical capacitive behavior," J. Mater. Chem. A, vol. 1, no. 21, pp. 6393-6399, 2013, doi: 10.1039/c3ta10433h.

[101] B. Senthilkumar, K. Vijaya Sankar, C. Sanjeeviraja, and R. Kalai Selvan, "Synthesis and physico-chemical property evaluation of PANI-NiFe 2O4 nanocomposite as electrodes for supercapacitors," J. Alloys Compd., vol. 553, pp. 350-357, 2013, doi: 10.1016/j.jallcom.2012.11.122.

Chapter 5

Nanoferrites-Based Drug Delivery Systems as Adjuvant Therapy for Cancer Treatments. Current Challenges and Future Perspectives

Felipe Ocampo Osorio, Jhon Augusto Jativa Herrera, Oscar Moscoso Londoño and César Leandro Londoño Calderón

Abstract

Cancer is the second cause of death worldwide, whose treatment often involves chemotherapy. In a conventional therapy, drug is transported (and usually absorbed) across biological membranes through diffusion and systemic transport. The pathway that medicine must travel before reaching the desired location, can bring adverse or unwanted effects, which are mainly the result of: low bioavailability, low solubility and toxicity. To avoiding risks, nanoparticles coated with the drug could be used as a therapeutic substance to selectively reach an area of interest to act without affecting non-target cells, organs, or tissues (drug delivery). Here, the goal is to enhance the concentration of the chemotherapeutic drug in the disease parts of the body. Among all nanostructured systems, ferrites attract worldwide attention in drug delivery applications. It is due to their versatile magnetic and physicochemical properties. Here, it is reviewed and analyzed recent advances in synthesis, morphology, size, magnetic properties, functionalization with a focus in drug delivery applications of nanoferrites.

Keywords: Ferrites, Nanostructures, Functionalization, Drug-loading, Drug delivery, Cancer

1. Introduction

Cancer is a disease originating from unregulated cell growth. Those cells can spread throughout the body, causing erroneous behaviors in organs or tissues [1]. Cancer is one of the principal problems in public health and currently is the second leading cause of death worldwide. According to the American Joint Committee on Cancer (AJCC) and the International Union for Cancer Control (UICC), there are many types of cancer treatments. Treatment or therapy depends on the cancer, as well as its stage of progress. Most individuals with cancer receive a combination of treatments, such as surgery with chemotherapy and/or radiation therapy [2]. Chemotherapy is crucial in the prevention of tumor recurrence and progression. Some patients have been treated with chemotherapeutic agents (e.g., Doxorubicin, Cyclodextrin, Cisplatin, Taxanes, Gemcitabine, among others) for long-term survival.

Despite recent advances in treatments for various types of cancer, the recurrence rate and severe side effects still are a problem. To improve the life quality of cancer patients, more efficient and accurate targeting treatment is an urgent need.

A nanotechnology-based drug delivery system may provide a feasible means to solve the previous challenges. This kind of technology can be a formulation or device that enables a therapeutic substance to selectively reach an area of interest to act without affecting non-target cells, organs, or tissues [3]. One of the most studied, promising, and simplest ways to transport pharmaceutical compounds in the body is using nano entities as delivery vehicles [4]. Moreover, some nanoparticulated systems exhibit sensitivity to external stimuli, such as visible light, near-infrared light, ultrasound, AC or DC magnetic fields, among others. These stimuli could be made use as a tool to flexible control of dose magnitude and timing from the responsiveness (triggered remotely) [5].

Specifically, nanoferrites have been attracted worldwide scientific community attention for applications against cancer due to [6]:

- Their great potential for hyperthermia treatments.

- The possibility to guide nanoparticles to specific regions using an external magnetic field.

- The chance to remotely activate the drug release in a controlled way (alternating magnetic fields).

Ferrites are compounds derived from iron oxides, whose composition allows tuning the magnetic properties. According to the magnetic atoms disposition and its chemical environment, material can manifest hard of soft magnetic properties. Nanostructure ferrites have received the most attention for drug delivery applications due to their versatile magnetic and chemical properties [7].

Previous reports have pay attention to the use of nanoferrites in biomedical applications for:

- Improving magnetic resonance imaging sensitivity [8].

- Effective targeted treatment of lung cancer [7].

- Magnetite nanoparticles as an advanced platform for cancer theranostics [9].

- Hydrogel beads-based nanoferrites in novel drug delivery platforms [10].

- Magnetic and superparamagnetic ferrites for cancer therapy applications [11].

- Iron oxide and substitute ferrite nanoparticles in drug delivery [12].

- The toxicity of spinel ferrite nanoparticles [13].

- Biosensing platform on ferrite nanoparticles [14].

Thus, the chapter aims to correlate the morphology, size, ferrite type, magnetic properties, functionalization, and pharmacokinetics. These correlations allow obtaining a perspective to the physical targeting precision for cancer drug delivery applications. Furthermore, we also discussed the current challenges and future perspectives of nanoferrites in the field of oncology.

2. Size, morphology, and magnetic properties of nanoferrites for drug delivery in cancer

The size, morphology, and magnetic properties of nanoparticles in drug delivery applications, have been identified as keys parameters in the literature [1, 2]. The easy way to tune these properties is from the synthetic routes [15]. The growth mechanism involved in the final morphology and structure is not completely clear. The conditions synthesis and their correlation with the physicochemical properties have been discussed in the literature [16]. Here, we focus on the recent advances in morphology, size, magnetic properties, and their relationship with the synthetic routes of nanoferrites used in drug delivery for cancer. **Table A1** shows a summary of these properties recently reported in the literature. For there, it is clear that the synthetic routes more employed for nanoferrites synthesis are:

- *Chemical coprecipitation:* it is a straightforward and inexpensive method. In this case, the precursor salts solutions containing the cationic metals are mix into an alkaline medium in a stoichiometric proportion.

- *Hydrothermal:* here, the chemical reactions take place in aqueous solutions at pressure and temperature higher than the room conditions.

- *Sol–gel*: the chemical reactions of hydrolysis and condensation are carried out of precursors in solution.

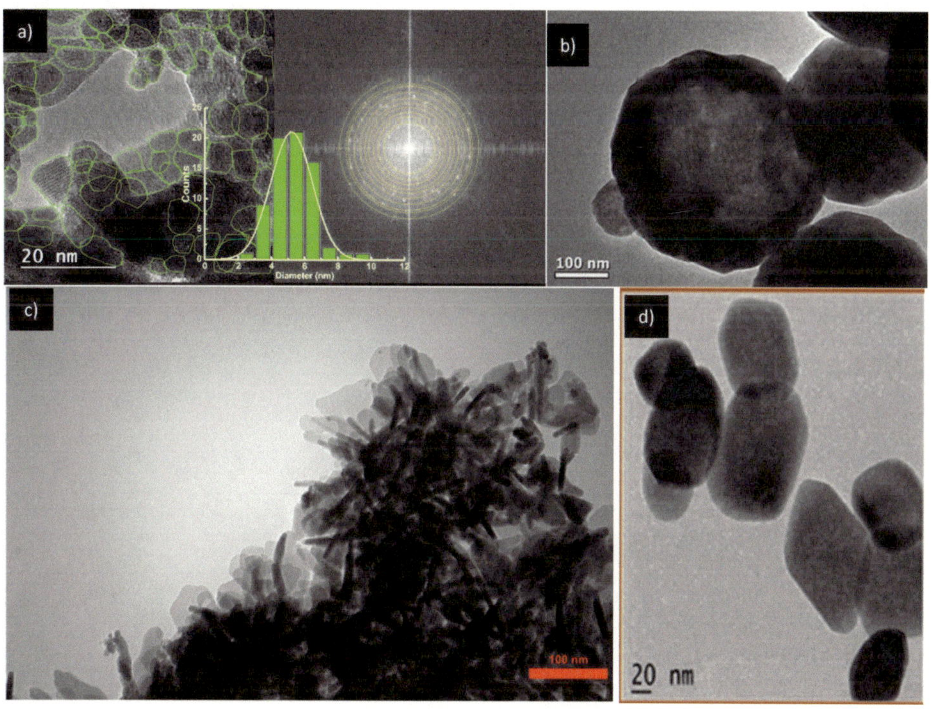

Figure 1.
Transmission Electron microscopy (TEM) images for a) calcium ferrite nanoparticles with a size of 5 nm, reproduced from Ref. [17] with permission of the editors, b) magnetite hollow spheres of diameter ~350 nm, reproduced from Ref. [65] with permission of the editors, c) magnetite nanorods, reproduced from Ref. [48] with permission of the editors, and d) magnetite hexagonal nanoparticles, reproduced from Ref. [56] with permission of the editors. Copyright 2019 MDPI, 2016 Nature, 2020 Elsevier, and 2021 the Royal Society of Chemistry.

- *Solvothermal*: it is like the hydrothermal technique. The difference is the use of nonaqueous solutions.
- Less popular synthesis techniques for nanoferrites obtention are thermal decomposition, sonochemical, thermal treatment, and thermolysis.

Tripathy et al. [14] reported a comparison among the different techniques for ferrite nanoparticles obtention.

Based on some scientific reports, nanoferrites used in cancer drug delivery applications range from 5.2 *nm* to 300 *nm* (**Table A1**). Calcium ferrite ($CaFe_2O_4$) obtained by coprecipitation is the smallest nanostructure system. Magnetite (Fe_3O_4) fabricate from the solvothermal method is the larger one (**Figure 1**). However, particles larger than 200 *nm* segregate by mechanical filtering and eventually get removed by phagocytic cells. Nanoparticles with sizes smaller than 10 nm lead to renal filtration and accumulation into the fenestration of the kidneys' glomerular endothelium. Therefore, the most effective drug delivery agents possess sizes ranging between 10 and 100 nm [68]. However, Sivaraj et al. [69] suggest that the nuclear membrane pores allow entry of nanoparticles with a size below 9 *nm*. Nanoparticles penetration into the cells may be maximized by surface functionalization with small molecules (e.g. folate, proteins, peptides, antibodies, and aptamers). This penetration induces receptor-mediated endocytosis, caveolae-mediated endocytosis, lipid raft mediated endocytosis, and/or micropinocytosis. After endocytosis in cancer therapy, nanomaterial releases maximum drug to inhibit the DNA/RNA synthesis and mitochondria damage.

The most common ferrite nanoparticles use for cancer drug delivery systems ranging from 20 *nm* to 30 *nm* (**Table A1**). Moreover, the most popular morphology

FT	Method	S	M_S	H_C	M_R	Reference
Fe_3O_4	Coprecipitation	11	59	0	—	[20]
$CoFe_2O_4$	Thermal decomposition	13	70.7	—	30.2	[22]
$MnFe_2O_4$	Sonochemical	13	34.9	0	0	[23]
Mn-Zn (Fe_2O_4)	Coprecipitation	15	56.0	0	0	[25]
$NiFe_2O_4$	Solvothermal	17	70	0	0	[29]
Fe_3O_4	Thermal treatment	23	7.1	143.8	2.2	[32]
Fe_3O_4	Coprecipitation	30	47.6	0	3.8	[39]
Fe_3O_4	Coprecipitation	35	36.3	0	—	[40]
Fe_3O_4	Coprecipitation	40	1.57	69.1	0.15	[41]
$CoFe_2O_4$	Coprecipitation	43	36.02	0	0	[42]
$GdFe_2O_4$	Coprecipitation	90	47	0	0	[57]
$CoFe_2O_4$	Solvothermal	104	51.8	0	0	[59]
$CaFe_2O_4$	Sol–gel	112	14.9	—	0.38	[61]
$MnFe_2O_4$	Coprecipitation	140	56.1	42.6	5.2	[62]
$CoFe_2O_4$	Thermolysis	200	51.1	0	0	[64]
$CoFe_2O_4$	Coprecipitation	250	40	1.7	—	[65]

Table 1.
Summary of spherical nanoparticles ferrite type (FT) *obtained by different methods with their sizes* (S *in* nm), *saturation magnetization* (M_S *in* emu/g), *coercivity* (H_C *in* Oe), *and remanence* (M_R *in* emu/g), *reported in the literature. All the magnetic properties were reported at room temperature.*

obtained from the synthetic routes is spherical particles (**Table 1**). Nanorods and particles with hexagonal shapes are the less common nanostructures used for drug delivery in cancer applications (**Figure 1**).

A complete understanding of magnetic properties is essential for a proper implementation of nanoferrites in drug delivery applications [6]. The saturation magnetization (M_s), coercive force (H_C), and remanence (M_R) are the most popular magnetic parameters reported for nanoferrites to cancer drug delivery applications (**Table A1**). Nanoferrites with the highest magnetic response (M_s) are cobalt ferrite ($CoFe_2O_4$) with a size of 15 nm obtained by sonochemical technique [28]. The smallest saturation magnetization was reported to zinc ferrite ($ZnFe_2O_4$) nanostructures (75 nm), which were synthesized by the sol–gel method [53]. Usually, nanoferrites used in cancer drug delivery applications show superparamagnetic behavior. Superparamagnetic nanoparticles evidence zero coercivity and remanence at temperatures above the blocking one (**Table 1**). In other cases, ferrite nanostructures with coercivity as high as 3409 Oe are used in cancer drug delivery applications ($CoFe_2O_4$ nanofibers with a diameter of 50 nm [45]. Moreover, cobalt ferrite nanoparticles show the highest magnetic remanence of 30.2 emu/g with a size of 30 nm obtained by thermal decomposition (**Table 1**).

3. Nanoferrites functionalized and functional groups for drug delivery in cancer

Non-functionalized nanoferrites (non-coating material on their surface) seems to be not optimal for drug delivery application. Surface energy minimization processes can promote agglomeration, percolation as well as other unwanted effects. Some of the most common problems whit this kind of nanosystems are [12]:

1. Agglomeration due to the attractive forces leading to non-stability of the nanoparticle dispersion.

2. Toxicity represents a problem in bare nanoferrites when they use without functionalization.

3. Bare nanoparticles do not have a functional group on their surface. This makes it hard to link drugs molecules.

To deal with these problems, nanoparticles have been coating with organic or inorganic molecules (functionalization). The surface engineering of ferrites could be accomplished during nanoparticle synthesis (in-situ) or after this (ex-situ). A detailed review of the coating and functionalization strategies was reported for nanoparticles in drug delivery applications by Pinelli et al. [70]. The surface functionalization procedure and choice of appropriate solvent are crucial factors for obtaining nanoferrites. Here, the repulsive interactions among nanoparticles prevent agglomerations [71]. Moreover, functionalization promotes several advantages such as stable dispersions, biocompatibility, biodegradability, and reduced toxicity. Usually, functionalized nanoparticles loaded with drugs adopt covalent/noncovalent interaction methods. Conjugation of a drug to a carrier by non-biodegradable linkages results in: changing the drug chemical units, reducing drug efficacy, and displaying relevant side effects. The drug remains unharmed by using physical adsorption for drug conjugation, and no changes occur in the chemical units and the controlled drug release behavior. In this case, the idea deals with functionalized nanoparticles that have an opposite electrical charge to the cancer drug to promote the electrostatic interaction [42]. Moreover, surface functionality gives significant strength to bind and adsorb cancer drugs using specific functional groups. The characterization techniques for studying the functional groups attached to nanoparticles for drug delivery applications have been reported previously [12].

Some examples of functional groups commonly used to functionalized nanoferrites in drug delivery applications are:

- The carboxyl functional group of the meso, 2–3-dimercaptosuccinic acid (DMSA), was used to functionalize cobalt ferrite nanoparticles [22].
- Magnetite nanoparticles functionalized with mesoporous silica (SiO_2) [58].
- Zinc ferrite nanoparticles coated with hydroxyapatite as an intermediate of the cancer drug [72].
- The carboxyl functional groups of citrate molecules use to functionalize manganese ferrite nanorods [24].
- Calcium ferrite functionalized with biomolecules (casein). The hybrid molecule combines the merits of both inorganic and organic counterparts [61].

The functionalization can allows high drug encapsulation, stabilizes the nanocarrier, and reaches the cancer site-specific. Furthermore, the coating uses to reach the target cells without getting removed by the reticuloendothelial system of the body and to have a capable surface for keeping the drug unharmed until reaching the location of interest. The performance enhancement achieves through functionalization with suitable ligands that will bind to the aimed receptors of pathological tissues. The size of the nanocarrier has paramount importance for rendering it absorbable by tumor tissues [68]. The inclusion of active targeting functionalities results in drug accumulation within tumors, tumor cells, or immune cells and allows for reduced dosages due to specificity. Functionalized ferrite nanoparticles have been used for: a) imitate ligand binding to receptors, b) for initiation of cellular signaling, c) for increased stimulation of immune cells to better infiltrate and extinguish immunosuppressive tumors [73]. Commonly, the pH of cancer cells (tumor) is acidic ranging between 4 and 5. It is due to the presence of lactic acid, which starts due to inefficient consumption of glucose [74]. On the other hand, the pH in an extracellular matrix or bloodstream is natural (pH = 7) [75]. This difference in pH offers to fabricate functionalized nanoparticles as a pH-sensitive trigger for drug delivery applications.

The most popular drugs for cancer delivery applications, using ferrites as nanocarriers are: Doxorubicin [58], 5-Fluorouracil [21], Docetaxel [76], Hesperidin and Eugenol [60], Curcumin [77], Tamoxifen [55], Cisplatin [78], Nilotinib [79], Camptothecin [38], and Telmisartan [20]. Hydrophobicity of the orally administered drugs for cancer treatments has low systemic bioavailability [80]. It produces low water solubility and can cause serious adverse effects [62].

Among functionalized nanoferrites investigated to load cancer drugs, one can find:

1. Zinc ferrite functionalized with Polyethylene Glycol (PEG) and chitosan loaded with Curcumin [80]. Chitosan takes cationic amine functional groups, at low pH, which would involve an ionic gelation process with polyanions to form nanoparticles. It is used as an effective drug carrier, where the reactive amine groups on the chitosan side chain are used for functional group modifications. The hydrophobically modified chitosan improves the encapsulation efficiency of the carrier towards the hydrophobic drugs [34].

2. Cobalt ferrite nanoparticles functionalized with DMSA used the amine functional group of Doxorubicin molecules. Here, it is attached through electrostatic interaction and/or hydrogen bonding interactions with the carboxylic functional group of the DMSA [22].

3. Magnetite nanoparticles functionalized with mesoporous silica used the amine group of Doxorubicin to attach [58].

4. Zinc ferrite nanoparticles functionalized with hydroxyapatite had covalent bonds with the zoledronic acid drug. Amino or hydroxyl functional groups presented in hydroxyapatite are a strong chemical bond with the mineral material of bone phases [54].

5. Magnetite nanoparticles functionalized with gelatin using the functional groups -NH^{3+}-. It produces by partial hydrolysis of collagen to interact with Doxorubicin [81].

6. Calcium ferrite nanoparticles functionalized with biomolecules (casein), which allows the conjugation of targeting ligands with functional groups. Actively bind with specific receptors that may be overexpressed on tumor cells, allowing improved biodistribution and delivery of the drugs at the cancer site [61].

7. Manganese ferrite nanorods functionalized with citrate molecules to electrostatically attach Doxorubicin [24].

Proteins are promising carriers for drug delivery applications. The main advantages are the abundance of active sites, improved biocompatibility, easy availability, and pH-dependent swelling behavior. The last one allows the programmed release of the cytotoxic agent in response to the acidic cancer microenvironment [82].

DFT calculations demonstrated Cisplatin on graphene oxide can be adsorbed by the functionalized nanoferrites. Here, hydrogen bonds forming with hydroxyl and epoxy functional groups. It involves the formation of the amide bond between Cisplatin and the COOH functional group of graphene oxide. In the case of glutaraldehyde, the functional group is CHO, which formed the amide bond between Cisplatin and thc CHO functional group [18].

4. Drugs loaded on functionalized nanoferrites for cancer treatments

Drug-loading of nanoparticles plays an essential role in drug delivery systems. There are several ways through which the drug can load with the functionalized nanoparticle:

- *Encapsulation*. It can entrap inside the nanoparticle.

- *Functionalization*. It can coat the surface of the nanoparticle.

- *Chemically linked*. It can be bond with the functionalized particle itself.

The second key point in functionalized nanoparticles design is the necessity to provide the nanoparticles with specific properties. The interaction with the external environment in the human body increases the targeting action towards determined sites [70].

Drug-loading involves several variables such as the solvent type and amount of it, the temperature, time of loading, and the drug-loading capacity. The most popular solvent for drug-loading is water (see **Table A2**). Less popular solvents involved in drug-loading are ethanol, dichloromethane, and saline solution. Usually, the solvent quantity varies from 1 *ml* to 200 *ml*. The drug-loading capacity represents the amount of drug loader per unit weight of the nanoparticle. Drug-loading represents the percentage of the nanoparticle mass that is due to the encapsulated drug. Loading capacity can calculate by the amount of total entrapped drug divided by the total nanoparticle weight. The drug-loading values reported for nanoferrites ranging from 0.016 [64] to 3.3 [63]. These values correspond to cobalt ferrite loaded with Doxorubicin and Docetaxel, respectively. The loading-drug temperature ranges from 4°C [54] to 55°C [33].

From **Table A2** many reports did not include the drug-loading solvent, the solvent quantity, and the drug loading capacity. The efficiency of drug-loading measure by a high-performance liquid chromatography system (HPLC) [30] or ultraviolet–visible spectroscopy (UV–Vis) [18]:

$$Drug - loading\% = \frac{total\ amount\ of\ drug - free\ amount\ of\ drug}{total\ amount\ of\ drug} x100 \qquad (1)$$

The free amount of the drug is measure by the absorbance of the supernatant in a UV–Vis spectrophotometer at the maximum wavelength of the dissolved drug. The nanoferrites can magnetically remove from the solution instead of the centrifugation process. The maximum wavelength for anticancer drugs are: Doxorubicin at 479 *nm* [23], Curcumin at 425 *nm* [34], Camptothecin at 480 *nm*, [68], 5-Fluorouracil at 266 *nm* [21], Cisplatin at 300 *nm* [18], Imanitib at 260 *nm* [31], Telmisartan at 296 *nm* [20], and Tamoxifen at 250 *nm* [55].

The time of loading is one of the essential factors in drug-loading. **Figure 2** shows a summary of the drug-loading efficiency results reported in the literature. The highest efficiency for drug-loading (98,3%) is reporting for calcium ferrite loaded with Curcumin in ethanol solvent at 100 *mL* with a drug-loading capacity of 0.4, at room temperature for 3 *h* [34]. The lowest efficiency for drug-loading (8,4%) is reporting for cobalt ferrite. It is loaded with Docetaxel in 10 *mL* of dichloromethane at room temperature for 1 *h*.

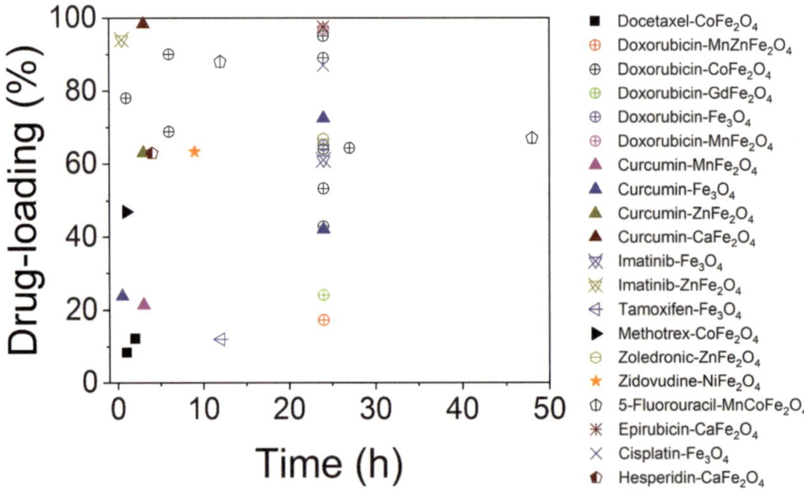

Figure 2.
Summary of the drug-loading percentage as a function of the time reported in the literature. All the data plotted are shown in ***Table A2****.*

Other alternatives for drug-loading of nanoferrites composites include:

1. Anchored nanoferrites of cobalt [28] and manganese [23] on graphene oxide were developed for controlled drug delivery nanocomposites.

2. Doxorubicin and nickel ferrite nanoparticles were incorporate into N-carboxymethyl chitosan/poly(ε-caprolactone) nanofibers for drug delivery applications [51].

5. Drug delivery of functionalized nanoferrites for cancer treatments

Conventional drug delivery methods rely on the absorption of drugs and transport across biological membranes through diffusion and systemic transport. The targeted drug delivery, on the contrary, focuses on enhancing the concentration of the chemotherapeutic drug in the disease parts of the body [87]. The drug release studies, usually, are realized in simulated physiological conditions and measured by HPLC [30] or UV–Vis spectra [88]. For UV–vis spectrophotometer, the percentage of release drug is given by [49]:

$$Drug\ release\% = \frac{amount\ of\ release\ drug}{amount\ of\ loaded\ drug} x100 \tag{2}$$

The drug release mechanisms evaluate with different models, such as zero-order, first-order, Higuchi, Korsmeyer-Peppas, and Hixson-Crowell. A detailed explanation of these five mathematical models to investigate drug release kinetic on *in-vitro* release data was reported by Jafari *et al.* [75]. The best mathematical model with a high correlation coefficient determines the suitable mathematical model and confirms drug release kinetics. Some results reported in the literature are summarized next:

- Curcumin drug loading on vanillin-chitosan coated with calcium ferrite hybrid nanoparticles as a carrier [34]. In most cases, the release mechanism follows non-Fickian diffusion, which may be due to the porous nature of the material, swelling ability, or the presence of an excess amount of surface adsorbed drug on the nanoparticles.

- Doxorubicin hydrochloride and Methotrexate drugs load on magnetite nanoparticles based on polyurethane matrices. The best fitting for the drug's release was the Higuchi kinetic model [75].

- Hesperidin drug loaded on magnetic casein-$CaFe_2O_4$ nanohybrid carrier conjugated with progesterone. Here, the release profile exhibiting the best fit towards the Higuchi model. Fickiand diffusion was validated as the release mechanism, which is a concentration gradient process [61].

- Doxorubicin drug loaded on carboxymethyl chitosan/poly(ε-caprolactone)/doxorubicin/nickel ferrite core-shell fibers. Here, the Korsmeyer-Peppas model showed the best pharmacokinetic fit [51].

- Hesperidin and Eugenol drugs loaded on folic acid functionalized BSA-$CaFe_2O_4$ nanohybrid carrier. The Korsmeyer-Peppas model showed the best fit for releasing the drug. The release mechanism at pH 1.2 is by anomalous diffusion. It is a combination of Fickian diffusion and the gradual erosion of the polymer. The release data for pH 5.8 and pH 7.4 fits well with the Higuchi model indicating a surface diffusion mechanism, in other words, the diffusion of the surface-bound [60].

- Cisplatin drug loaded on magnetite nanocomposite. The results of the kinetic studies suggest that the most proper model to interpret the release of the drug in pH 5.5 is the Korsmeyer-Peppas equation. The value of the release exponent in this model suggests that the prime mechanism of drug release is diffusion and Fickian [78].

- Doxorubicin drug loaded on pure and lanthanum doped bismuth ferrite nanostructures. The kinetic studies and adsorption isotherms revealed that the adsorption of the drug fitted well to the pseudo-second-order and Freundlich isotherm models. The adsorption of doxorubicin followed the multi-layered heterogeneous adsorption. The probable loading mechanism was electrostatic interaction [89].

The efficiency *in-vitro* tumor-targeted drug delivery of the nanoferrites loaded with anticancer drugs is evaluated by fluorescence microscopy imaging [58]. Here, the authors used human cancer cell lines as: MCF-7, A-549 [81], A431 [37], SKOV-3, MDA-MB-231 [61], SK-BR3 [33], MDA-MB-231, MCF-10A [90] in a culture media which is incubate in presence of the nanoferrites. Moreover, cytotoxicity determines the efficiency of the formulation [51]. The cytotoxic effect of the nanoformulation tests the cell viability (MTT) assay. It evaluated the ability of viable cells to reduce MTT to formazan crystals. The following equation used to calculate the % of cell viability is [91]:

$$\%Cell\ viability = \frac{average\ sample\ read}{average\ control\ read} x100 \qquad (3)$$

In-vivo antitumor therapy came tests in mice. Here, hepatoma cell lines (H22) inoculate into the back of the hind leg through subcutaneous injection. When the size of the tumor was grown to about 40 mm^3 were treated with the drug-loading nanoferrites. All the formulations were injected intravenously through the tail of mice. The tumor inhibition rates could be determined by fluorescence microscopy [67]:

$$Tumor\ inhibition\ rates = 1 - \frac{tumor\ volume\ with\ drug\ group}{tumor\ volume\ in\ control\ group} x100 \qquad (4)$$

Nanoferrites have got importance in terms of biological applications due to their physicochemical properties. To enhance their cancer therapeutic effect stimuli-responsive combine treatments have been developed:

- *Magnetic hyperthermia therapy.* Here, drug release may be activated applying an external alternating magnetic field which transforms electromagnetic energy into heat and induces the drug carrier to release its contents into the target site [34, 36, 57]. The combined techniques can enhance the therapeutic effect by increasing the blood flow and improving the oxygen supply to the tumor sites when increasing the temperature of the tumor sites from 37°*C* to hyperthermia temperature 42–45°C. This phenomenon can also enhance the drug delivery efficacy and increase the drug dosage to the target tumor sites. To determine the intracellular drug delivery of ferrite nanoparticles load with anticancer drugs, which were placed in dialysis membrane tubes and dialyzed at 37°*C* with different pHs [22].

- *Chemo-sonodynamic therapy*. It is another strategy for cancer treatment because the low-intensity ultrasound caused the activation of drug-loaded magnetic nanosonosensitizers. With synthesized ultrasound-sensitive nanocarriers, chemo-sonodynamic therapy is a generator of cellular reactive oxygen species, mitochondrial damage, and inducer effect through the release of the loaded drug in magnetic nanoferrite [53].

- *Photodynamic therapy*. Photodynamics is a method to treat cancer via light and photosensitizing chemical material. Here, small bandgap energy of the nanocarrier is desired to excite the electrons by light. The electrons transferred from the conduction band to the valence band produce an electron–hole pair. These pairs react with H_2O and O_2 and produce reactive oxygen species (ROS) [77].

- *Microwave irradiation*. Microwave irradiation added to magnetite nanocomposite increases the drug release [19].

Usually, drug delivery is dramatically pH-dependent. Most of the papers reported in the literature studied the influence of pH on the release behavior of the carrier. pH variations at different physiological situations trigger a controlled delivery of drugs at different sites. The pH-responsive drug release under three conditions of simulated gastric fluid (pH 1.2), cancer microenvironment (pH 5.4), and simulated body fluid (pH 7.4) during a determined time [61]. **Table A3** shows the influence of the pH on the release efficiency of the carrier. From there, in all cases, the acidic pH stimuli the rate of drug delivery. **Table 2** shows the drug release percentage at cancer microenvironment conditions for nanoferrite formulations. The time of drug release is one of the essential factors in drug delivery. **Figure 3**

System	pH	t (h)	T (°C)	DR (%)	Reference
Fe_3O_4– Doxorubicin	5.5	80	37	60	[58]
$Mg_{0.5}Co_{0.5}$ Fe_2O_4-5-fluorouracil	4.5	48	37	97	[21]
$CoFe_2O_4$ -Doxorubicin	5.4	75	37	42	[83]
$CoFe_2O_4$ -Doxorubicin	5.4	120	37	52	[47]
$CoFe_2O_4$-Hesperidin and eugenol	5.8	24	35	73.7	[60]
$MnFe_2O_4$ -curcumin	5.5	120	37	90	[60]
$CoFe_2O_4$ - Doxorubicin	4.0	24	37	60	[43]
$MnFe_2O_4$- Doxorubicin	5,5	10	37	17.16	[24]
Fe_3O_4 - Curcumin	5.0	120	37	40	[92]
Fe_3O_4 -Telmisartan	5.5	52	37	82	[20]
$ZnFe_2O_4$ - Curcumin	5.5	96	37	64.71	[53]
Fe_3O_4- Cisplatin	5,5	48	37	96	[78]
Fe_3O_4-Doxorubicin	5.8	72	25	70	[26]
$CoFe_2O_4$-Doxorubicin	5.4	72	37	80	[42]

Table 2.
Summary of drug delivery conditions and results reported in the literature for ferrite nanoparticles loaded with anticancer drugs (system). The main conditions are the cancer microenvironments (pH), the time (t), and the temperature (T) of release. The drug release (DR) percentage measures the efficiency of the process.

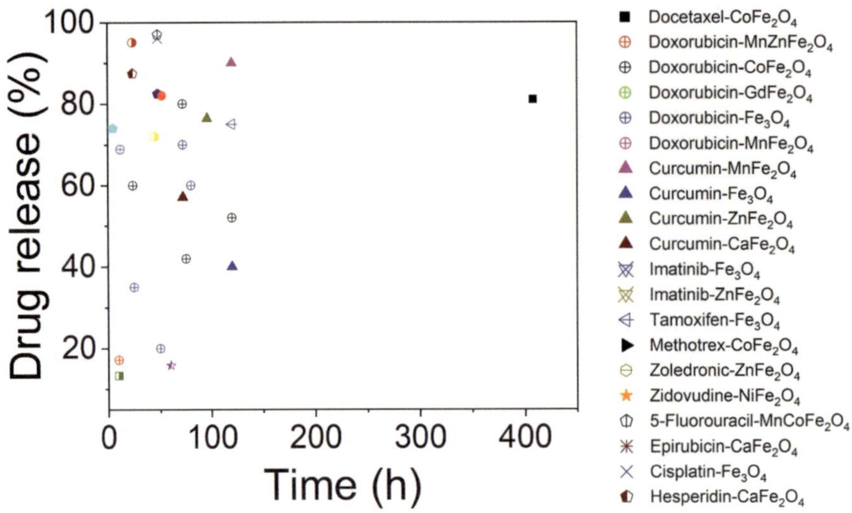

Figure 3.
*Summary of the drug delivery efficiency as a function of the time reported in the literature. All the data plotted are shown in **Table** 2 and **A**3.*

shows a summary of the drug delivery efficiency results reported in the literature. The highest efficiency for drug delivery (97%) is reporting for magnesium-cobalt ferrite loaded with 5-fluorouracil for 48 h [21]. The lowest efficiency for drug delivery (8,9%) was reported for magnetite load with Curcumin for 37 h.

6. Conclusions

Recent advances reviewed on synthetic routes for the obtention of nanoferrites for drug delivery applications. The most popular ferrite is magnetite obtained by chemical coprecipitation method with sizes ranging from 20 nm to 30 nm, and spherical shape. Moreover, it reviews the magnetic properties of ferrite nanoparticles. Often, the nanoferrites are superparamagnetic. Coated the nanoparticle's surface with organic or inorganic molecules makes the nanostructures optimal for drug delivery applications. Functionalization reduces the agglomeration and toxicity of the nanoferrites. Physical adsorption among the functional groups of the cancer drugs and the coated molecules on the nanoparticles preserve the chemical structure of the medicament. Oncology drugs were detailed for drug delivery applications. The most popular solvent for drug-loading is water. It discussed the influence of parameters such as: pH, temperature, and time on drug-loading.

It reviewed the main drug release mechanisms for investigating pharmacokinetics. The release mechanism is highly dependent on the pH, the type of drug, and the nanocarrier. It discussed the stimuli-responsive combine treatments for cancer drug delivery applications. Some challenges persist:

- The nonspecific accumulation of the drug and the lack of real-time monitoring of the delivery.

- Develop novel multifunctional theranostic platforms with the abilities of intelligent controlled released and in-vivo site targeting delivery and treatment of illnesses.

- Improve the delivery effectiveness of a drug by maintaining the concentration of the drug between the effective and toxic levels.
- Inhibiting the dilution of the drug in the body fluids, and allowing targeting and localization of a drug at a specific site
- Determine the optimal temperature and concentration of drug required to promote effective apoptosis.

Acknowledgements

C.L.L.C and J.A.J.H acknowledge Ministerio de Ciencia Tecnología e Innovación - Colombia (MINCIENCIAS, Call # 848). O.M.L and C.L.L.C acknowledge Unidad de Investigación from Universidad Autónoma de Manizales (Colombia).

Conflict of interest

The authors declare no conflict of interest.

Appendices and nomenclature

FT	Method	S (nm)	Mo	M_S (emu/g)	H_C (Oe)	M_R (emu/g)	R
Ca	Coprecipitation	5	Spherical	12.8	3.4	0.06	[17]
Fe	Coprecipitation	7	Spherical	—	—	—	[18]
Mn	Coprecipitation	10	Semi- spherical	13	12.9	0.15	[19]
Fe	Coprecipitation	11	Spherical	59	0	—	[20]
MnCo	Glycol-Thermal	12	Spherical	—	—	—	[21]
Co	Thermal decomposition	13	Spherical	70.7*	—	30.2	[22]
Mn	Sonochemical	13	Spherical	34.9	0	0	[23]
Mn	Hydrolysis	15	Nanorod	69.3	116	—	[24]
MnZn	Coprecipitation	15	Spherical	56	0	0	[25]
Fe	Coprecipitation	15	Spherical	—	—	—	[26]
Co	Coprecipitation	15	Spherical	—	—	—	[27]
Co	Sonochemical	15	Spherical	94	0	0	[28]
Ni	Solvothermal	17	Spherical	70	0	0	[29]
Ho	Hydrothermal	21	Spherical	47.8	—	—	[30]
Zn	Hydrothermal	21	Coral	2	0	—	[31]
Fe	Thermal treatment	23	Spherical	7.11	143.9	2.21	[32]
Ni	Sol–Gel	24	Spherical	—	—	—	[33]
Ca	Combustion of solutions	25	Spherical	9.11	—	0.02	[34]

FT	Method	S (nm)	Mo	M_S (emu/g)	H_C (Oe)	M_R (emu/g)	R
Co	Coprecipitation	26	Spherical	49	549	—	[35]
Co	Microwave heating	27	Spherical	90.5	830	—	[36]
Ca	Coprecipitation	30	Semi- spherical	58.3	—	—	[37]
Fe	Hydrothermal	30	Hexagonal	1.2	175	0.91	[38]
Fe	Precipitation	30	Spherical	47.6	0	3.81	[39]
Fe	Coprecipitation	35	Semi- spherical	36.3	0	—	[40]
Fe	Coprecipitation	40	Spherical	1.57	69.1	0.15	[41]
Co	Coprecipitation	42	Spherical	36.0	0	0	[42]
Co	Solvothermal	43	Spherical	47.7	—	—	[43]
Co	Coprecipitation	43	Spherical	62.4	—	—	[44]
Co	Precipitation	50	Nanofiber	3.9	3409	2.1	[45]
Co	Coprecipitation	50	Spherical	59.9	—	—	[46]
Co	Sol–Gel	50	Semi- spherical	64	—	—	[47]
Fe	Hydrothermal	50	Nanorods	22.2	0	—	[48]
Fe	Coprecipitation inverse	60	Spherical	63	—	—	[49]
Ca	Combustion of solutions	60	Spherical	9.1	—	0.02	[50]
Ni	Sol–Gel	70	Spherical	50.5	—	—	[51]
Ho	Hydrothermal	74	Semi- spherical	43	—	—	[52]
Zn	Sol–Gel	75	Spherical	0.6	65.6	3.8	[53]
Zn	Coprecipitation	80	Semi- spherical	31	100	—	[54]
Fe	Coprecipitation	80	Semi- spherical	80.1	—	—	[55]
Fe	Hydrothermal	90	Hexagonal	34	714	—	[56]
Gd	Coprecipitation	90.1	Spherical	47	0	0	[57]
Fe	Solvothermal	95	Spherical	59	—	—	[58]
Co	Solvothermal	104	Semi- spherical	51.8	0	0	[59]
Ca	Sol–Gel	112	Spherical	15	0.16	0.38	[60]
Ca	Sol–Gel	112	Spherical	15	—	0.38	[61]
Mn	Coprecipitation	140	Spherical	56.1	42.6	5.22	[62]
Co	Thermolysis	157	Spherical	13.7	0	—	[30]
Fe	Hydrothermal	200	Spherical	71.9	0	0	[63]
Co	Thermolysis	200	Spherical	51.1	0	0	[64]
Co	Coprecipitation	250	Spherical	40	1.7	—	[65]
Mn	Coprecipitation	300	Nanorods	18	—	—	[66]
Fe	Solvothermal	300	Spherical	57.4	57.5	—	[67]

Table A1.
Summary of nanoparticles ferrite type (FT) obtained by different methods with their sizes (S), morphology (Mo), saturation magnetization (M_S), coercivity (H_C) and remanence (M_R), reported in the literature (R). The magnetic properties were reported at room temperature.

FT	Drug	Sol	QS	LD	T(°C)	t(h)	%L	R
Co	D	D	10	—	RT	1	8,4	[30]
Fe	T	E	5	0.2	30	12	12	[55]
Co	D	W	30	3.3	RT	2	12,2	[63]
Ni	C	E	—	—	RT	12	—	[29]
Co	Do	W	5	—	—	1	—	[65]
Co	Do	W	25	0.04	25	24	—	[83]
Mn-Zn	Do	W	10	—	RT	24	17.3	[25]
Mn	C	W	25	—	RT	3	21,3	[84]
Fe	C	—	—	0.2	RT	0,5	23,7	[85]
Gd	Do	S	5	0.2	RT	24	24	[57]
Fe	C	W	10	0.2	RT	24	42	[48]
Co	Do	—	—	—	RT	24	43	[59]
Co	M	W	—	0.005	RT	1	47	[35]
Fe	C	W	25	0.3	RT	—	52,3	[32]
Co	Do	—	—	0.67	25	24	53,3	[42]
Fe	I	W	30	—	RT	24	61	[82]
Ca	H	E	—	—	RT	4	62,9	[40]
Zn	C	E	1	—	—	3	63	[60]
Ni	Z	W	20	—	55	9	63.2	[80]
Co	Do	—	—	—	RT	24	64	[66]
Co	Do	W	1	—	25	27	64.3	[33]
Co	Do	W	5	0.016	30	24	65	[43]
Fe	Do	—	—	0.5	RT	24	65	[27]
Zn	Zo	—	—	—	4	24	66,7	[64]
Mn-Co	5-F	—	—	—	RT	48	67	[58]
Co	Do	W	5	—	29.85	6	68,8	[54]
Fe	C	W	50	—	RT	24	72,4	[21]
Zn	C	W	50	0.016	RT	—	76,4	[71]
Co	Do	W	5	—	RT	1	78	[41]
Ca	Ci	—	—	—	25	—	86.5	[53]
Fe	Cis	—	—	—	RT	24	87	[86]
Mn	5-F	—	—	0.2	RT	12	88	[82]
Co	Do	—	—	—	RT	24	89	[18]
Co	Do	—	—	0.2	RT	6	90	[62]
Zn	I	—	—	0.42	45	0,5	94,0	[47]
Co	Do	S	200	—	37	24	95	[22]
Mn	Do	W	10	0.5	RT	24	97	[31]
Ca	E	—	—	—	37	24	97.5	[37]
Ca	C	E	100	0.4	RT	3	98,3	[34]

Table A2.
Summary of nanoparticles ferrite type (FT) loaded (Fe: Magnetite) with different anticancer drugs (D: Docetaxel, T: Tamoxifen, C: Curcumin; do: Doxorubicin, M: Methotrexate, I: Imatinib, H: Hesperidin, Z: Zidovudine, zo:Zoledronic acid, 5-F: 5-fluorouracil, ci: Cinnamaldehyde, Cis: Cisplatin, and E: Epirubicin,), the solvent (sol) used for loading (D: Dichloromethane, E: Ethanol, W: Water, and S: Saline solution) and the quantity (QS, in mL), the loading capacity (LD), the temperature (T), time (t, in hours), and percentage (% L) of loading, with their references (R).

System	pH	t (h)	T (°C)	DR (%)	Reference
Fe_3O_4 - Doxorubicin	7.4	80	37	20	[58]
$Mg_{0.5}Co_{0.5}$ Fe_2O_4–5-fluorouracil	6.5 7.4	48	37	73 73	[21]
Fe_3O_4- Docetaxel	7.4	48	37	82.43	[76]
$CoFe_2O_4$ -Doxorubicin	7.4	120	37	22	[47]
$CoFe_2O_4$- Hesperidin and Eugenol	1.2 7.4	24	35	87.44 54.29	[60]
$MnFe_2O_4$ - curcumin	7.4	120	37	41	[84]
$CoFe_2O_4$ - Doxorubicin	7.4	50	37.5	42.6	[43]
$CoFe_2O_4$ - Doxorubicin	7.4	24	37	30	[59]
$MnFe_2O_4$ - Doxorubicin	7.4	10	37	11.93	[24]
$CoFe_2O_4$ - Curcumin	7.4	72	37	57,1	[77]
Fe_3O_4- Curcumin	7.4	120	37	8,9	[92]
$CoFe_2O_4$ - Docetaxel	7.4	408	37	81	[30]
Fe_3O_4 -Telmisartan	7.4	52	37	25	[20]
$ZnFe_2O_4$ - Curcumin	7.4	96	37	76.45	[53]
Fe_3O_4 - Tamoxifen	7.4	120	37	75	[55]
$Ag_{(1-x)}NiFe_2O_4$ - Curcumin	6	5	31	74	[29]
$ZnFe_2O_4$ - Curcumin	6	15	38	—	[93]
$NiFe_2O_4$ -, $CoFe_2O_4$ - and Fe3O4 -Doxorubicin	7.4	0.0069	37	13.3	[71]
Fe_3O_4 - Cisplatin	7.4	48	37	93	[78]
Fe_3O_4 - Doxorubicin	6,86	25	—	35	[39]
Fe_3O_4 - Doxorubicin	7	72	25	35	[26]
$CoGd_xFe_{2-x}O_4$ - Curcumin	7.4	24	30	95	[94]
$CoFe_2O_4$ - Doxorubicin	7.4	72	37	80	[27]
$ErFe_3O_4$- Camptothecin	7.4	45	37	72	[38]

Table A3.
Summary of drug delivery conditions and results, reported in the literature for ferrite nanoparticles loaded with anticancer drugs (system). The main conditions are the pH, the time (t) and the temperature (T) of release. The drug release (DR) percentage measure the efficiency of the process.

Author details

Felipe Ocampo Osorio, Jhon Augusto Jativa Herrera, Oscar Moscoso Londoño and César Leandro Londoño Calderón*
Physics and Mathematics Department, Faculty of Engineering, Autonomous University of Manizales, Manizales, Colombia

*Address all correspondence to: cesarl.londonoc@autonoma.edu.co

References

[1] Aisida SO, Akpa PA, Ahmad I, Zhao T, Maaza M, Ezema FI. Bio-inspired encapsulation and functionalization of iron oxide nanoparticles for biomedical applications. Eur Polym J. 2020 Jan;122: 109371.

[2] Shende P, Shah P. Carbohydrate-based magnetic nanocomposites for effective cancer treatment. Int J Biol Macromol. 2021 Apr;175:281–93.

[3] Pandey A. Role of Cyclodextrins in Nanoparticle-Based Systems for Drug Delivery. 2020;305–43.

[4] Maghsoudnia N, Eftekhari RB, Sohi AN, Zamzami A, Dorkoosh FA. Application of nano-based systems for drug delivery and targeting: a review. J Nanoparticle Res. 2020 Aug;22(8):245.

[5] Li F, Qin Y, Lee J, Liao H, Wang N, Davis TP, et al. Stimuli-responsive nano-assemblies for remotely controlled drug delivery. J Control Release. 2020 Jun;322:566–92.

[6] Soares PIP, Romão J, Matos R, Silva JC, Borges JP. Design and engineering of magneto-responsive devices for cancer theranostics: Nano to macro perspective. Prog Mater Sci. 2021 Feb;116:100742.

[7] Sadhasivam J, Sugumaran A. Magnetic nanocarriers: Emerging tool for the effective targeted treatment of lung cancer. J Drug Deliv Sci Technol. 2020 Feb;55:101493.

[8] Wang Y, Miao Y, Li G, Su M, Chen X, Zhang H, et al. Engineering ferrite nanoparticles with enhanced magnetic response for advanced biomedical applications. Mater Today Adv. 2020 Dec;8:100119.

[9] Zhao S, Yu X, Qian Y, Chen W, Shen J. Multifunctional magnetic iron oxide nanoparticles: an advanced platform for cancer theranostics. Theranostics. 2020;10(14):6278–309.

[10] Amiri M, Khazaeli P, Salehabadi A, Salavati-Niasari M. Hydrogel beads-based nanocomposites in novel drug delivery platforms: Recent trends and developments. Adv Colloid Interface Sci. 2021 Feb;288:102316.

[11] Avval ZM, Malekpour L, Raeisi F, Babapoor A, Mousavi SM, Hashemi SA, et al. Introduction of magnetic and supermagnetic nanoparticles in new approach of targeting drug delivery and cancer therapy application. Drug Metab Rev. 2020 Jan;52(1):157–84.

[12] Al-Rawi NN, Anwer BA, Al-Rawi NH, Uthman AT, Ahmed IS. Magnetism in drug delivery: The marvels of iron oxides and substituted ferrites nanoparticles. Saudi Pharm J. 2020 Jul;28(7):876–87.

[13] Kefeni KK, Msagati TAM, Nkambule TT, Mamba BB. Spinel ferrite nanoparticles and nanocomposites for biomedical applications and their toxicity. Mater Sci Eng C. 2020 Feb;107: 110314.

[14] Tripathy A, Nine MJ, Silva FS. Biosensing platform on ferrite magnetic nanoparticles: Synthesis, functionalization, mechanism and applications. Adv Colloid Interface Sci. 2021 Apr;290:102380.

[15] Ghosh N, Pant P, Bhuvaneswari S. Chemical Methodologies for Preparation of Micron and Nanometer Scale Ferrites - A Mini Review of Patents. Recent Pat Nanotechnol. 2008 Jan;2(1):8–18.

[16] Kharisov BI, Dias HVR, Kharissova O V. Mini-review: Ferrite nanoparticles in the catalysis. Arab J Chem. 2019 Nov; 12(7):1234–46.

[17] Pereira, Cardoso, Rodrigues, Amorim, Amaral, Almeida, et al.

Magnetoliposomes Containing Calcium Ferrite Nanoparticles for Applications in Breast Cancer Therapy. Pharmaceutics. 2019 Sep;11(9):477.

[18] Abdel-Bary AS, Tolan DA, Nassar MY, Taketsugu T, El-Nahas AM. Chitosan, magnetite, silicon dioxide, and graphene oxide nanocomposites: Synthesis, characterization, efficiency as cisplatin drug delivery, and DFT calculations. Int J Biol Macromol. 2020 Jul;154:621–33.

[19] Yang Z, Wang L, Liu Y, Liu S, Tang D, Meng L, et al. ZnO capped flower-like porous carbon-Fe3O4 composite as carrier for bi-triggered drug delivery. Mater Sci Eng C. 2020 Feb;107:110256.

[20] Dhavale RP, Dhavale RP, Sahoo SC, Kollu P, Jadhav SU, Patil PS, et al. Chitosan coated magnetic nanoparticles as carriers of anticancer drug Telmisartan: pH-responsive controlled drug release and cytotoxicity studies. J Phys Chem Solids. 2021 Jan;148:109749.

[21] Mngadi S, Mokhosi S, Singh M, Mdlalose W. Chitosan-Functionalized Mg0.5Co0.5Fe2O4 Magnetic Nanoparticles Enhance Delivery of 5-Fluorouracil In Vitro. Coatings [Internet]. 2020 May 2;10(5):446. Available from: https://www.mdpi.com/2079-6412/10/5/446

[22] Oh Y, Moorthy MS, Manivasagan P, Bharathiraja S, Oh J. Magnetic hyperthermia and pH-responsive effective drug delivery to the sub-cellular level of human breast cancer cells by modified CoFe2O4 nanoparticles. Biochimie. 2017 Feb;133: 7–19.

[23] Wang G, Ma Y, Zhang L, Mu J, Zhang Z, Zhang X, et al. Facile synthesis of manganese ferrite/graphene oxide nanocomposites for controlled targeted drug delivery. J Magn Magn Mater. 2016 Mar;401:647–50.

[24] R.K. C, Rajagopalan V, Sahu NK. Synthesis of manganese doped β-FeOOH and MnFe 2 O 4 nanorods for enhanced drug delivery and hyperthermia application. IET Nanobiotechnology. 2020 Dec;14(9): 823–9.

[25] Montha W, Maneeprakorn W, Tang I-M, Pon-On W. Hyperthermia evaluation and drug/protein-controlled release using alternating magnetic field stimuli-responsive Mn–Zn ferrite composite particles. RSC Adv. 2020;10 (66):40206–14.

[26] Nieciecka D, Celej J, Żuk M, Majkowska-Pilip A, Żelechowska-Matysiak K, Lis A, et al. Hybrid System for Local Drug Delivery and Magnetic Hyperthermia Based on SPIONs Loaded with Doxorubicin and Epirubicin. Pharmaceutics. 2021 Apr;13(4):480.

[27] Beagan AM, Alghamdi AA, Lahmadi SS, Halwani MA, Almeataq MS, Alhazaa AN, et al. Folic Acid-Terminated Poly(2-Diethyl Amino Ethyl Methacrylate) Brush-Gated Magnetic Mesoporous Nanoparticles as a Smart Drug Delivery System. Polymers (Basel). 2020 Dec;13(1):59.

[28] Wang G, Ma Y, Wei Z, Qi M. Development of multifunctional cobalt ferrite/graphene oxide nanocomposites for magnetic resonance imaging and controlled drug delivery. Chem Eng J [Internet]. 2016 Apr;289:150–60. Available from: http://dx.doi.org/10.1016/j.cej.2015.12.072

[29] Jose R, J R, Jothi NSN. The synthesis and characterisation of curcumin loaded Ag (1-X) Ni X Fe 2 O 4 for drug delivery. Mater Technol. 2021 May;36 (6):339–46.

[30] Panda J, Satapathy BS, Mandal B, Sen R, Mukherjee B, Sarkar R, et al. Anticancer potential of docetaxel-loaded cobalt ferrite nanocarrier: an in vitro study on MCF-7 and MDA-MB-

231 cell lines. J Microencapsul. 2021 Jan; 38(1):36–46.

[31] Amiri M, Gholami T, Amiri O, Pardakhti A, Ahmadi M, Akbari A, et al. The magnetic inorganic-organic nanocomposite based on ZnFe2O4-Imatinib-liposome for biomedical applications, in vivo and in vitro study. J Alloys Compd [Internet]. 2020 Dec;849:156604. Available from: https://linkinghub.elsevier.com/retrieve/pii/S0925838820329686

[32] Ehsan N, Magmoud N, Heidar TR, Reza ZE, Gholamreza F, Ramezan AT. In vivo and In vitro Biocompatibility Study of Fe3O4@ZnO and Fe3O4@SiO2 as Photosensitizer for Targeted Breast Cancer Drug Delivery. J Sci. 2020; 31(4):12.

[33] Joshy KS, Augustine R, Mayeen A, Alex SM, Hasan A, Thomas S, et al. NiFe 2 O 4 /poly(ethylene glycol)/lipid–polymer hybrid nanoparticles for anti-cancer drug delivery. New J Chem. 2020;44(42):18162–72.

[34] Kamaraj S, Palanisamy UM, Kadhar Mohamed MSB, Gangasalam A, Maria GA, Kandasamy R. Curcumin drug delivery by vanillin-chitosan coated with calcium ferrite hybrid nanoparticles as carrier. Eur J Pharm Sci [Internet]. 2018 Apr;116:48–60. Available from: https://linkinghub.elsevier.com/retrieve/pii/S0928098718300307

[35] Shahzad K, Mushtaq S, Rizwan M, Khalid W, Atif M, Din FU, et al. Field-controlled magnetoelectric core-shell CoFe2O4@BaTiO3 nanoparticles as effective drug carriers and drug release in vitro. Mater Sci Eng C. 2021 Feb;119: 111444.

[36] Radmansouri M, Bahmani E, Sarikhani E, Rahmani K, Sharifianjazi F, Irani M. Doxorubicin hydrochloride - Loaded electrospun chitosan/cobalt ferrite/titanium oxide nanofibers for hyperthermic tumor cell treatment and controlled drug release. Int J Biol Macromol. 2018 Sep;116:378–84.

[37] Shahabadi N, Razlansari M, Khorshidi A, Zhaleh H. Investigation of controlled release properties and anticancer effect of folic acid conjugated magnetic core–shell nanoparticles as a dual responsive drug delivery system on A-549 and A-431 cancer cell lines. Res Chem Intermed. 2020 Sep;46(9): 4257–78.

[38] Kaliyamoorthi K, Sumohan Pillai A, Alexander A, Ramasamy S, Arivarasu A, Enoch IVM V. Designed poly(ethylene glycol) conjugate-erbium-doped magnetic nanoparticle hybrid carrier: enhanced activity of anticancer drug. J Mater Sci. 2021 Feb;56(5):3925–34.

[39] Ren M-X, Wang Y-Q, Lei B-Y, Yang X-X, Hou Y-L, Meng W-J, et al. Magnetite nanoparticles anchored on graphene oxide loaded with doxorubicin hydrochloride for magnetic hyperthermia therapy. Ceram Int. 2021 Jul;47(14):20686–92.

[40] Karimi Ghezeli Z, Hekmati M, Veisi H. Synthesis of Imatinib-loaded chitosan-modified magnetic nanoparticles as an anti-cancer agent for pH responsive targeted drug delivery. Appl Organomet Chem. 2019 Apr;33(4): e4833.

[41] Rostami M, Aghajanzadeh M, Zamani M, Manjili HK, Danafar H. Sono-chemical synthesis and characterization of Fe3O4@mTiO2-GO nanocarriers for dual-targeted colon drug delivery. Res Chem Intermed. 2018 Mar;44(3):1889–904.

[42] Arkaban H, Khajeh Ebrahimi A, Yarahmadi A, Zarrintaj P, Barani M. Development of a multifunctional system based on CoFe 2 O 4 @polyacrylic acid NPs conjugated to folic acid and loaded with doxorubicin

for cancer theranostics. Nanotechnology. 2021 Jul;32(30): 305101.

[43] Zhang H, Wang J, Zeng Y, Wang G, Han S, Yang Z, et al. Leucine-coated cobalt ferrite nanoparticles: Synthesis, characterization and potential biomedical applications for drug delivery. Phys Lett A. 2020 Aug;384 (24):126600.

[44] Amiri M, Akbari A, Ahmadi M, Pardakhti A, Salavati-Niasari M. Synthesis and in vitro evaluation of a novel magnetic drug delivery system; proecological method for the preparation of CoFe2O4 nanostructures. J Mol Liq. 2018 Jan;249:1151–60.

[45] Sangeetha K, Vidhya G, Girija EK, Ashok M. Fabrications of magnetic responsive hydroxyapatite platform: In vitro release of chemo drug for cancer therapy. Mater Today Proc. 2020;26: 3579–82.

[46] Hassanzadeh-Tabrizi SA, Norbakhsh H, Pournajaf R, Tayebi M. Synthesis of mesoporous cobalt ferrite/ hydroxyapatite core-shell nanocomposite for magnetic hyperthermia and drug release applications. Ceram Int. 2021 Jul;47(13): 18167–76.

[47] Ghanbari M, Davar F, Shalan AE. Effect of rosemary extract on the microstructure, phase evolution, and magnetic behavior of cobalt ferrite nanoparticles and its application on anti-cancer drug delivery. Ceram Int. 2021 Apr;47(7):9409–17.

[48] Kermanian M, Naghibi M, Sadighian S. One-pot hydrothermal synthesis of a magnetic hydroxyapatite nanocomposite for MR imaging and pH-Sensitive drug delivery applications. Heliyon. 2020 Sep;6(9):e04928.

[49] Pooresmaeil M, Namazi H. β-Cyclodextrin grafted magnetic graphene oxide applicable as cancer drug delivery agent: Synthesis and characterization. Mater Chem Phys. 2018 Oct;218:62–9.

[50] Bilas R, Sriram K, Maheswari PU, Sheriffa Begum KMM. Highly biocompatible chitosan with super paramagnetic calcium ferrite (CaFe2O4) nanoparticle for the release of ampicillin. Int J Biol Macromol. 2017 Apr;97:513–25.

[51] Abasalta M, Asefnejad A, Khorasani MT, Saadatabadi AR. Fabrication of carboxymethyl chitosan/ poly(ε-caprolactone)/doxorubicin/ nickel ferrite core-shell fibers for controlled release of doxorubicin against breast cancer. Carbohydr Polym. 2021 Apr;257:117631.

[52] Kaliyamoorthi K, Ramasamy S, Pillai AS, Alexander A, Arivarasu A, Enoch IVMV. Camptothecin-loaded holmium ferrite nanocarrier. Expanded activity on breast cancer cells. Mater Lett. 2021 Feb;285:129164.

[53] Hafezi M, Rostami M, Hosseini A, Rahimi-Nasrabadi M, Fasihi-Ramandi M, Badiei A, et al. Cur-loaded ZnFe2O4@mZnO@N-GQDs biocompatible nano-carriers for smart and controlled targeted drug delivery with pH-triggered and ultrasound irradiation. J Mol Liq. 2021 Jan;322: 114875.

[54] Seyfoori A, Ebrahimi SAS, Omidian S, Naghib SM. Multifunctional magnetic ZnFe2O4-hydroxyapatite nanocomposite particles for local anti-cancer drug delivery and bacterial infection inhibition: An in vitro study. J Taiwan Inst Chem Eng. 2019 Mar;96: 503–8.

[55] Theragnostic Magnetic Core-Shell Nanoparticle as Versatile Nanoplatform for Magnetic Resonance Imaging and Drug Delivery. Biointerface Res Appl Chem [Internet]. 2021 Feb 8;11(5): 13276–89. Available from: https://b

iointerfaceresearch.com/wp-content/uploads/2021/02/20695837115.1327613289.pdf

[56] Ramasamy S, Dhamecha D, Kaliyamoorthi K, Pillai AS, Alexander A, Dhanaraj P, et al. Magnetic hydroxyapatite nanomaterial–cyclodextrin tethered polymer hybrids as anticancer drug carriers. Mater Adv. 2021;2(10):3315–27.

[57] Mekonnen TW, Birhan YS, Andrgie AT, Hanurry EY, Darge HF, Chou H-Y, et al. Encapsulation of gadolinium ferrite nanoparticle in generation 4.5 poly(amidoamine) dendrimer for cancer theranostics applications using low frequency alternating magnetic field. Colloids Surfaces B Biointerfaces. 2019 Dec;184: 110531.

[58] Fang Z, Li X, Xu Z, Du F, Wang W, Shi R, et al. Hyaluronic acid-modified mesoporous silica-coated superparamagnetic Fe3O4 nanoparticles for targeted drug delivery. Int J Nanomedicine. 2019 Jul;Volume 14: 5785–97.

[59] Shi Z, Zeng Y, Chen X, Zhou F, Zheng L, Wang G, et al. Mesoporous superparamagnetic cobalt ferrite nanoclusters: Synthesis, characterization and application in drug delivery. J Magn Magn Mater. 2020 Mar; 498:166222.

[60] Uma Maheswari P, Muthappa R, Bindhya KP, Meera Sheriffa Begum KM. Evaluation of folic acid functionalized BSA-CaFe2O4 nanohybrid carrier for the controlled delivery of natural cytotoxic drugs hesperidin and eugenol. J Drug Deliv Sci Technol. 2021 Feb;61: 102105.

[61] K. Purushothaman B, P UM, K. M. MSB. Magnetic casein-CaFe2O4 nanohybrid carrier conjugated with progesterone for enhanced cytotoxicity of citrus peel derived hesperidin drug towards breast and ovarian cancer. Int J Biol Macromol. 2020 May;151: 293–304.

[62] Karimi Z, Abbasi S, Shokrollahi H, Yousefi G, Fahham M, Karimi L, et al. Pegylated and amphiphilic Chitosan coated manganese ferrite nanoparticles for pH-sensitive delivery of methotrexate: Synthesis and characterization. Mater Sci Eng C. 2017 Feb;71:504–11.

[63] Panda J, Satapathy BS, Majumder S, Sarkar R, Mukherjee B, Tudu B. Engineered polymeric iron oxide nanoparticles as potential drug carrier for targeted delivery of docetaxel to breast cancer cells. J Magn Magn Mater. 2019 Sep;485:165–73.

[64] Fan H, Li B, Shi Z, Zhao L, Wang K, Qiu D. A fibrous morphology silica-CoFe2O4 nanocarrier for anti-cancer drug delivery. Ceram Int. 2018 Feb;44 (2):2345–50.

[65] Mandal Goswami M. Synthesis of Micelles Guided Magnetite (Fe3O4) Hollow Spheres and their application for AC Magnetic Field Responsive Drug Release. Sci Rep. 2016 Dec;6(1):35721.

[66] Abbasi Pour S, Shaterian HR, Afradi M, Yazdani-Elah-Abadi A. Carboxymethyl cellulose (CMC)-loaded Co-Cu doped manganese ferrite nanorods as a new dual-modal simultaneous contrast agent for magnetic resonance imaging and nanocarrier for drug delivery system. J Magn Magn Mater. 2017 Sep; 438:85–94.

[67] Liu X, Wang C, Wang X, Tian C, Shen Y, Zhu M. A dual-targeting Fe3O4@C/ZnO-DOX-FA nanoplatform with pH-responsive drug release and synergetic chemo-photothermal antitumor in vitro and in vivo. Mater Sci Eng C. 2021 Jan;118:111455.

[68] Kaliyamoorthy K, Pillai AS, Alexander A, Arivarasu A, Enoch IVMV, Ramasamy S. β-Cyclodextrin-folate functionalized poly(lactic-co-glycolide)–superparamagnetic ytterbium ferrite hybrid nanocarrier for targeted delivery of camptothecin. Mater Sci Eng C. 2021 Mar;122:111796.

[69] Mehnath S, Das AK, Verma SK, Jeyaraj M. Biosynthesized/green-synthesized nanomaterials as potential vehicles for delivery of antibiotics/drugs. In 2021. p. 363–432.

[70] Pinelli F, Perale G, Rossi F. Coating and Functionalization Strategies for Nanogels and Nanoparticles for Selective Drug Delivery. Gels. 2020 Feb;6(1):6.

[71] Mannu R, Karthikeyan V, Velu N, Arumugam C, Roy VAL, Gopalan A-I, et al. Polyethylene Glycol Coated Magnetic Nanoparticles: Hybrid Nanofluid Formulation, Properties and Drug Delivery Prospects. Nanomaterials. 2021 Feb;11(2):440.

[72] Seyfoori A, Ebrahimi SAS, Omidian S, Naghib SM. Multifunctional magnetic ZnFe2O4-hydroxyapatite nanocomposite particles for local anti-cancer drug delivery and bacterial infection inhibition: An in vitro study. J Taiwan Inst Chem Eng [Internet]. 2019 Mar;96:503–8. Available from: https://linkinghub.elsevier.com/retrieve/pii/S1876107018305777

[73] Day NB, Wixson WC, Shields CW. Magnetic systems for cancer immunotherapy. Acta Pharm Sin B. 2021 Apr;

[74] Mandal P, Panja S, Banerjee SL, Ghorai SK, Maji S, Maiti TK, et al. Magnetic particle anchored reduction and pH responsive nanogel for enhanced intracellular drug delivery. Eur Polym J. 2020 Apr;129:109638.

[75] Jafari S, Soleimani M, Badinezhad M. Application of different mathematical models for further investigation of in vitro drug release mechanisms based on magnetic nano-composite. Polym Bull. 2021 Jan;

[76] Song C, Gao C, Zhao J, Wang Z. Construction of long-circulation EpCAM targeted drug delivery system and its application in the diagnosis and treatment of breast cancer. J Biomater Appl. 2021 Mar;35(8):947–57.

[77] Naderi E, Aghajanzadeh M, Zamani M, Sharafi A, Naseri M, Danafar H. The Effect of Calcination Temperature on the Anticancer Activity of CaFe2O4@PVA Nanocarriers: Photodynamic Therapy and Drug Delivery Study. J Inorg Organomet Polym Mater. 2020 Dec;30(12):5261–9.

[78] Siavashy S, Soltani M, Ghorbani-Bidkorbeh F, Fallah N, Farnam G, Mortazavi SA, et al. Microfluidic platform for synthesis and optimization of chitosan-coated magnetic nanoparticles in cisplatin delivery. Carbohydr Polym. 2021 Aug;265:118027.

[79] Zhalechin M, Dehaghi SM, Najafi M, Moghimi A. Magnetic polymeric core-shell as a carrier for gradual release in vitro test drug delivery. Heliyon. 2021 May;7(5):e06652.

[80] Sawant VJ, Bamane SR, Shejwal RV, Patil SB. Comparison of drug delivery potentials of surface functionalized cobalt and zinc ferrite nanohybrids for curcumin in to MCF-7 breast cancer cells. J Magn Magn Mater. 2016 Nov;417:222–9.

[81] Huang C-H, Chuang T-J, Ke C-J, Yao C-H. Doxorubicin–Gelatin/Fe3O4–Alginate Dual-Layer Magnetic Nanoparticles as Targeted Anticancer Drug Delivery Vehicles. Polymers (Basel). 2020 Aug;12(8):1747.

[82] Purushothaman BK, Maheswari P U, Sheriffa Begum K M M. <scp>pH</scp> and magnetic field responsive protein-inorganic nanohybrid conjugated with biotin: A biocompatible carrier system targeting lung cancer cells. J Appl Polym Sci. 2021 Mar;138(10):49949.

[83] Cheraghi A, Davar F, Homayoonfal M, Hojjati-Najafabadi A. Effect of lemon juice on microstructure, phase changes, and magnetic performance of CoFe2O4 nanoparticles and their use on release of anti-cancer drugs. Ceram Int. 2021 Jul;47(14):20210–9.

[84] Aghajanzadeh M, Naderi E, Zamani M, Sharafi A, Naseri M, Danafar H. In vivo and in vitro biocompatibility study of MnFe 2 O 4 and Cr 2 Fe 6 O 12 as photosensitizer for photodynamic therapy and drug delivery of anti-cancer drugs. Drug Dev Ind Pharm. 2020 May;46(5):846–51.

[85] Ayyanaar S, Kesavan MP, Sivaraman G, Maddiboyina B, Annaraj J, Rajesh J, et al. A novel curcumin-loaded PLGA micromagnetic composite system for controlled and pH-responsive drug delivery. Colloids Surfaces A Physicochem Eng Asp. 2019 Jul;573:188–95.

[86] Dey C, Baishya K, Ghosh A, Goswami MM, Ghosh A, Mandal K. Improvement of drug delivery by hyperthermia treatment using magnetic cubic cobalt ferrite nanoparticles. J Magn Magn Mater. 2017 Apr;427:168–74.

[87] Kaliyamoorthy K, Pillai AS, Alexander A, Arivarasu A, Enoch IVMV, Ramasamy S. β-Cyclodextrin-folate functionalized poly(lactic-co-glycolide)–superparamagnetic ytterbium ferrite hybrid nanocarrier for targeted delivery of camptothecin. Mater Sci Eng C [Internet]. 2021 Mar;122:111796. Available from: https://linkinghub.elsevier.com/retrieve/pii/S0928493120337152

[88] Ehi-Eromosele CO, Ita BI, Iweala EEJ. Silica coated LSMO magnetic nanoparticles for the pH-Responsive delivery of 5-Fluorouracil anticancer drug. Colloids Surfaces A Physicochem Eng Asp. 2017 Oct;530:164–71.

[89] Abbasi MA, Ali Z, Qamar Z, Shahzad K, Siddiqui HK, Atif M, et al. Phase pure synthesis of lanthanum doped bismuth ferrite nanostructures for the adsorption of doxorubicin. Ceram Int. 2021 May;47(10):14390–8.

[90] Klein S, Distel LVR, Neuhuber W, Kryschi C. Caffeic Acid, Quercetin and 5-Fluorocytidine-Functionalized Au-Fe3O4 Nanoheterodimers for X-ray-Triggered Drug Delivery in Breast Tumor Spheroids. Nanomaterials. 2021 Apr;11(5):1167.

[91] Jermy R, Ravinayagam V, Alamoudi W, Almohazey D, Elanthikkal S, Dafalla H, et al. Tuning pH sensitive chitosan and cisplatin over spinel ferrite/silica nanocomposite for anticancer activity in MCF-7 cell line. J Drug Deliv Sci Technol. 2020 Jun;57:101711.

[92] Asgari M, Miri T, Soleymani M, Barati A. A novel method for in situ encapsulation of curcumin in magnetite-silica core-shell nanocomposites: A multifunctional platform for controlled drug delivery and magnetic hyperthermia therapy. J Mol Liq. 2021 Feb;324:114731.

[93] Jose R, J R, Jothi NSN. Synthesis and characterisation of stimuli-responsive drug delivery system using ZnFe 2 O 4

and Ag 1-X Zn x Fe 2 O 4 nanoparticles. Mater Technol. 2021 May;36(6):347–55.

[94] Javed F, Abbas MA, Asad MI, Ahmed N, Naseer N, Saleem H, et al. Gd3+ Doped CoFe2O4 Nanoparticles for Targeted Drug Delivery and Magnetic Resonance Imaging. Magnetochemistry. 2021 Mar;7(4):47.

Chapter 6

Development of Graphene Based Cobalt-Ferrites Nanocomposites for Microwave Shielding

Muhammad Siyar and Asghari Maqsood

Abstract

The study is related to cobalt ferrites nanocomposites embedded with graphene nanosheets, prepared by co-precipitation method. Various doping of graphene from 0.1% up to 1% were applied within the cobalt ferrite structure to study its microwave and mechanical effects on the nanocomposites. Microstructural analysis confirms the homogeneous dispersion and successful adhesion of graphene nanosheets within the cobalt ferrite matrix. Microwave absorbing capacity of these samples was studied by Agilent network analyzer in low frequency band of microwave (1MHz to 2 GHz), Results reveals that graphene incorporation not only improved the absorption capacity of cobalt ferrites (13dB-17d), but also widened its maximum absorption peak. This change was supposed to be due to inhomogeneity and combine effects of electric (graphene), and magnetic dielectric nature (cobalt ferrites). Further mechanical characterizations reveal that our composites samples have higher flexural strength (19.92 MPa for 1% loading) and improved toughness (>6000 J/mm2) compare to pure cobalt ferrites (10.28 MPa, 1000 J/mm2).

Keywords: Cobalt ferrites, nanocomposites, graphene, microwave shielding

1. Introduction

1.1 Introduction to ferrites

Ferrites are the magnetic mix metal oxides comprising the ferric ions as an essential constituent, while in mineralogy or in metallurgy the term ferrites refer to that material having a cubic crystal structure of spinal mineral. The ferrites application has been known from ancient times for multiple centuries. Magnetite or ferrous ferrite is a naturally occurring ferrite. These are a class of mix metal magnetic oxide materials where ferric ions are present as a basic constituent. In metallurgy or mineralogy any materials containing spinal mineral of cubic crystal structure is referred as ferrites. The Chinese found its use in the loadstone (i.e. Fe3O4) due to its weak permanent magnetism for navigation purposes as early in the 12th century [1, 2].

Hilbert a German scientist, in 1909 reported first systematic study on the relation between magnetic and chemical properties of many iron oxides, but faced hurdles to identify its magnetic phases. However near to 1930s modern ferrites were investigated by Forestier, Hilppert and Wille for their structural, magnetic and electrical properties. Japanese scientist's attempted to study magnetic oxides from 1930s to 1935, Snoek and co-workers were studying the ferrites in 1936; they found that one of the important properties of ferrites to use as core for inductors is the loss tangent per permeability, so called loss-factor (**Figure 1**) [3].

They also investigated that loss can always be minimized by introducing an air gap and the permeability not effected adequately. This finding, led Snoke to develope manganese-zinc ferrite, of low loss and ultimate permeability. It was 1945, that Snoke coined the physical and technological applications of modern ferrites and a new field of science came into being. Since that time ferrites brought stormy revolution in electrical and electronic areas, to improve its capability and endurance [4–6].

1.2 Structure of ferrites

Ferrites crystallize in three types of structures.

Garnet Ferrites: These are also cubic structural ferrites but with general formula Ln33 + Fe5O12, here Ln maybe, EU, Tb, Tm, Ho, Er, Gd or Ln [7].

Magnetoplumbit ferrites: With hexagonal structure, having general formula M2 + Fe12O19, where M may be Sr. or Ba [8].

Spinal ferrites: Having cubic crystalline structure with general formula M2 + Fe2O4, here M may be any metal such as Mn, Cd, Fe, Co, Cu, Mg or Al [9].

Spinal ferrites are the simplest structure, crystallizes among the three ferrites groups. The spinal ferrites unit cell is comprised mainly of oxygen lattice, having 32 oxygens, 16 trivalent irons along with 8 divalent metal ions, as shown in **Figure 2**. An interesting feature of oxygen lattice is its ions are arranged in face centered cubic structure, such that two types of interstitial sites are produced, at A site tetrahedral co-ordinate and octahedral B sites. The whole unit cell consists of two groups of four cubes. The positions of the ions are changing in these groups may share corners, faces or edges. Each group having four oxygens (Large spheres), located diagonally on each accountant and also on the tetrahedral corner. The tetrahedral oxygen ions surrounding the metal ions located in the center of the left hand side octant (small sphere, not shaded). This ion appeal to occupy A site of the intrinsic position produced by oxygen.

The second ionic group on right hand contains four metal ions enclosed by an octahedron resulted from six oxygen ions, and goes to fill the B site. Hence each unit cell consists of 64 A sites, only 8 of which are occupied and remaining unoccupied free sites. B sites are 32 in number, 16 of which are occupied [4, 5].

Figure 1.
Ferrites devices used in different applications [1]. It can be easily made in any design and shape, and hence give it more diversity to be applied in various magnetoelectric applications.

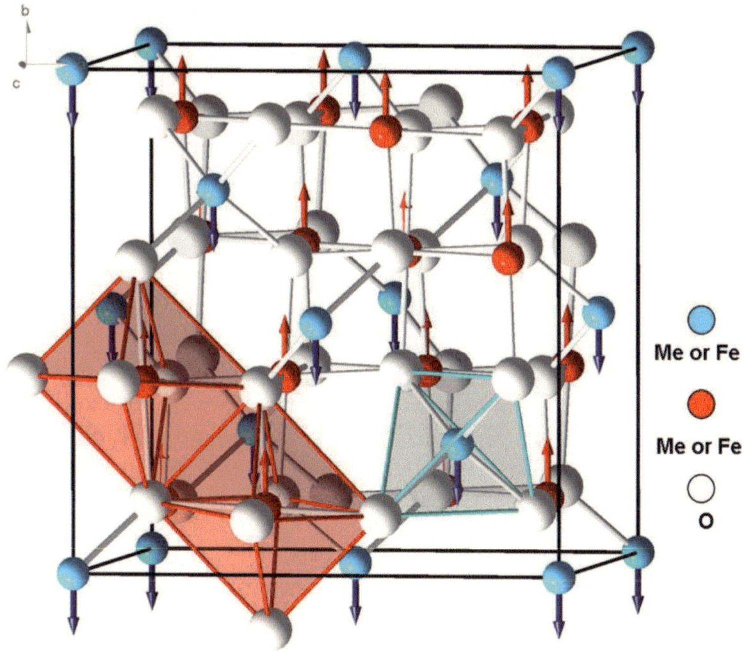

Figure 2.
Structure of spinal ferrite unit cell, comprised mainly of oxygen lattice, having 32 oxygens, 16 trivalent irons along with 8 divalent metal ions [5].

1.3 Synthesis routes of ferrites

As we are interested in ferrites composites, so we have to evaluate nano ferrites synthesis techniques, while coarse ferrites are produced by traditional ceramic processing. Nano particles ferrites are superior due to large surface area, very small size, high surface energy and excellent properties [9, 10].

Bottom –up and top-down approaches are reported so far, in bottom-up we take molecules or atoms to manipulate these in a systematic way to get the desired shape, size and properties of specific ferrite materials. In top-down approaches the bulk materials are broken down repeatedly until we get nano sized materials. Mechanical attrition done in high energy ball milling is an example of top-down approach.

Synthesis routes may be generally classified in two ways, the physical way (involving only physical changes such as phase, shape, and size of the matter) and chemical way (where permanent or chemical change occurs). Bottom-up techniques as lithography, ball milling and screen printing are all physical synthesis routes. Some chemical routes for nano ferrites synthesis are sol–gel, hydrothermal and co-precipitation reported from various laboratories [11, 12].

Maaz et.al reported the successful synthesis of cobalt ferrites nano particles via wet chemical route. In this study they have studied CoFe2O4 nanoparticles in the nanometers range size for low saturation magnetization and large coercivity. They succeeded to achieve an efficient control over the size of the synthesized cobalt ferrite particles via finely tuning the annealing temperature and time [13].

we have presented the synthesis of CoFe2O4 nanoparticles in the range 15–48 nm. The size of the nanoparticles was measured both by XRD and TEM and were in very good agreement with each other indicating that there was no agglomeration and that the size distribution of the prepared nanoparticles was small. The size of the nanoparticles appeared to increase linearly with annealing temperature and time most probably due to coalescence that increases with increasing temperature of anneal. It is evident that particle size and its distribution may be controlled both by controlling the rate of reaction and the annealing temperature and time. The very large coercivity and low saturation magnetization at 77 K in comparison with room temperature appear to be due to a pronounced growth of magnetic anisotropy at low temperatures. The observed magnetization remanence ratio of 0.47 at room temperature (very close to the value of 0.5 typical of a system of noninteracting single domain particles) suggests that CoFe2O4 nanoparticles exhibit an effective anisotropy that is uniaxial. The effective uniaxial anisotropy in magnetic nanoparticles has been explained as arising from surface effects that also lead to large anisotropy energy in nanoparticles. The coercivity shows a peak with particle size at a value much smaller than the single domain limit and is attributed to the onset of thermal effects at small enough particle sizes. We find that for smaller particles the saturation magnetization had a value that was significantly lower than the bulk value while the larger size particles have values approaching those of the bulk. The smaller value of MS in smaller particles is attributed to the greater fraction of surface spins in these particles that tend to be in a canted or spin glass like state with a smaller net moment.

1.4 Properties and applications

Ferrites are used in almost every electrical and electronic devices, it finds its use in the applications from simple lifting magnets to complex electrical and communication systems. Demand for ferrites is increasing rapidly due to the advancement in modern technology, such as satellite communications, radar systems, memory, and advance computer applications. The major upsurge for the passion of ferrites research is due to the development of modern small and fast power supplies, which are being used in cell phones, computers, communications, microprocessors, and entertainment applications [14].

Ferrites coined its modern use, first in a powerful application to enable the time dependent magnetic deflection of the electron array, in a television receiver. Mn-Zn ferrite emerged as an excellent candidate for core material in line-time-base and e.h.t transformer. Another excellent area of use was the field-shaping the yoke for beam deflection system in the picture tube. Later on, several ferrites' compositions i.e. Mg-Zn, Ni-Zn and Li-Zn etc. were applied for yokes application due to higher resistivity to enable direct winding of the coils on ferrite. However highly efficient TV deflection yokes are fabricated from Mn-Zn ferrite. Also, ferrite rods and plates make it possible to make the compact antennas for radio receivers (**Figure 3**) [12–14, 16].

The need for use of ferrites in line communication systems started, when the analog frequency divisions multiplex (FDM) systems demand for large numbers of ultimate performance, transformers and inductors operating in the frequency range from 40 kHz to 500 kHz. Ferrites were selected an excellent agent for core materials in these applications, due to a combination of high resistivity along with good magnetic nature. Their large frequency range along with ultimate resistivity successfully replaced the conventional metallic magnetic materials and further enabled FDMs to operate over wide and higher frequency range [17].

Other important applications of ferrites are summarized below.

- Flyback transformers [18]
- Fitter inductors [15]
- Magnetic amplifier [19]

Figure 3.
Digital TV receiver contains Mn-ferrites deflection yokes [15].

- Antenna cores
- IF transformers
- Magnetic switches and memories
- Magnetic drug delivery systems.

2. Materials and methods

Graphene based cobalt ferrites nano composites were prepared by facile co-precipitation route [20]. Homogeneous solution was first prepared by mixing appropriate amount of Fe (NO3)3.9H2O and 0.1 mol of Co (NO3)2.6H2O in deionized water followed by vigorous mixing. Appropriate amount of GO loading was applied to achieve 1%, 0.5% and 0.1% loading in respect to ferrite precursors. The prepared homogeneous mixture was stirred continuously and heated up to 95°C, while NaOH salt was added drop by drop. The process of mixing at the set temperature was continued for around 3 hours to obtain the dark black solution. After completion of reaction the solution was washed and filtered with the help of centrifuge again and again to get the neutral PH and remove the reaction media and impurities. Composite samples were dried in oven at 80 C for 24 hours and sintered for 3 hours at 700°C. Characterizations.

XRD spectra were recorded for all the samples in two thetas range from 10 to 70, on the pellet samples. Both the permeability and permittivity of our samples were measured by Agilent RF Impedance-Material analyzer (moel-E4991A), over the entire range from 1 MHz to 3GHz, with using 16453A test fixture for permittivity and 16454A for permeability on normal and torrid samples, respectively.

3. Results and discussion

3.1 Phase analysis of composites

Ferrite nano composites samples were analyzed by X-ray diffraction to study the structure and phases of the obtained samples. The obtained XRD spectra of all the samples maybe seen in the **Figure 4**. The corresponding planes (531), (440), (511), (422), (400), (222), (311), (220) and (111) have observed in all the samples, which confirm the successful formation of ferrite structure. Presence of graphene can be seen only in 1% graphene doped sample at 28 degrees two theta, while in other samples graphene amount were not enough to be reflected in the XRD spectra.

3.2 Electrical characterization

The microwave composed of two perpendicular magnetic and electrical vectors [21], as shown in **Figure 5**.

So efficient absorption can only be achieved by canceling both the electrical and magnetic components of it. When we use some media for microwave absorption, it should be capable of opposing both these perpendicular vectors. Here we want to study the absorption of microwave by incorporating electrically conducting graphene sheets into magnetic (cobalt ferrites) matrix.

3.3 Real permittivity and permeability

The behavior of frequency dependence of the real parts (ε′ and μ′) of permittivity and permeability for cobalt ferrites (S0) and its composites with different loading of graphene (S1, S2, S3) at room temperature is shown in **Figure 6**. As is clear from this study that all the samples exhibit frequency dependency for both permittivity and permeability in the low frequency range which become independent of frequency at higher frequency and shows consent behavior with respect to it. The variation of ε′ with respect to change in composition may be explained by Maxwell-Wagner theory of space charge polarization [22]. Due to the inhomogeneous structure composed of good conducting grains and poor conducting grain boundaries of dielectric materials

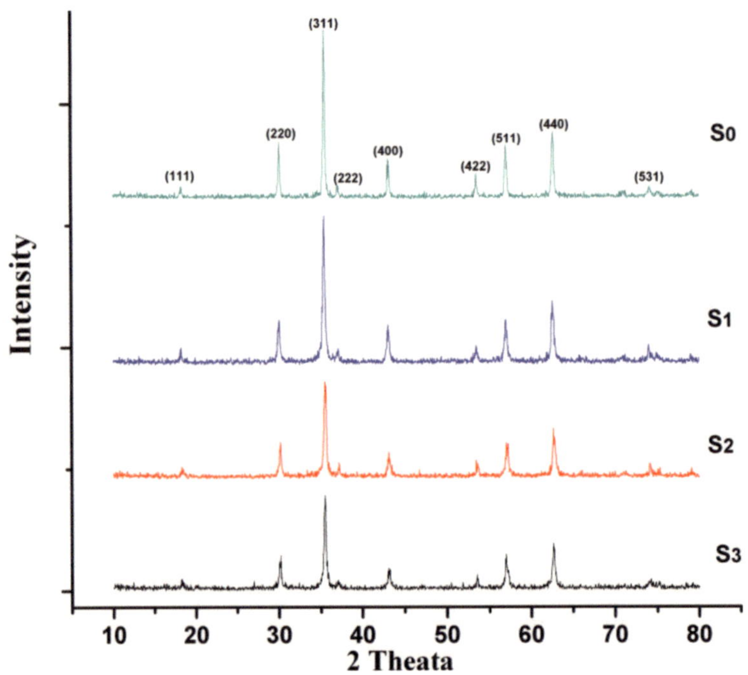

Figure 4.
XRD spectra of graphene based cobalt ferrites nano composites prepared by facile co-precipitation route. S_0 is pure ferrite while S1, S2 and S3 refers to the appropriate amount of GO loading of 1%, 0.5% and 0.1% in ferrite matrix.

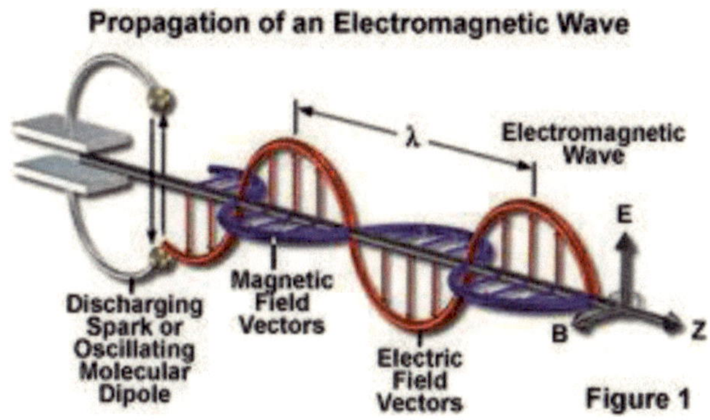

Figure 5.
Schematics for the demonstration of electromagnetic wave propagation and its various vectors.

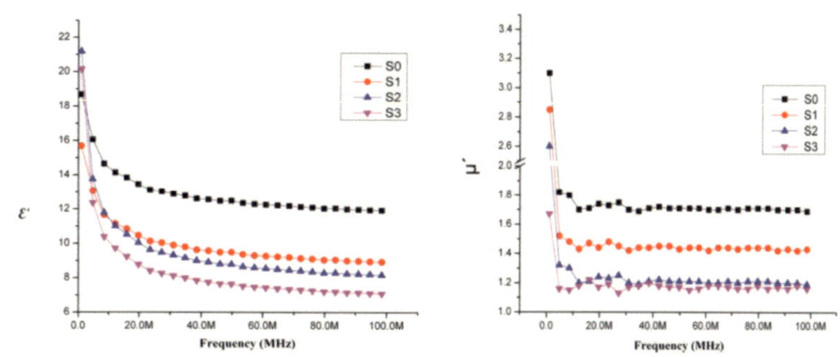

Figure 6.
Real part of permittivity and permeability of Co nanoferrite and composites samples in the frequency range of 1 MHz to 100 MHz.

space polarization occurs. In more detail the electrons are approaching to the grain boundaries passing through the conducting media within the grain by hopping, piled up along the boundary lines and produced polarization [23].

When the applied field frequency is increased the probability of electron approaching towards grain boundaries decreases and hence dielectric constant also goes down. For composite samples the lower curves are due to the conducting graphene sheets in between the cobalt ferrite particles, which tend to decrease polarization due to conduction of charges along these conducting sheets.

3.4 Complex permittivity and permeability

Complex spectra for permittivity and permeability are taken for all the composites and pure cobalt ferrites samples in the frequency range up to 100 MHz are shown in **Figure 7**. Both imaginary parts of permittivity and permeability refer to the energy loss from the material to the electric and magnetic field respectively [24]. For permittivity the complex portion is highly dependent on the frequency of all the four samples and decrease rapidly with increasing frequency which becomes less dependent after 50 MHz if we compare the response of cobalt ferrites to its composite samples, it can be seen from the **Figure 6**. That at the start composites samples has higher loss than the pure ferrites while going at higher frequency the loss became higher for ferrite sample comparable to all other composites. Although there is some gap in the response of ferrites and composite samples but it is negligible and hence we can assume that no major affect is there in ferrites loss to electric field due to graphene incorporation. This decrease may be due to the presence of comparatively

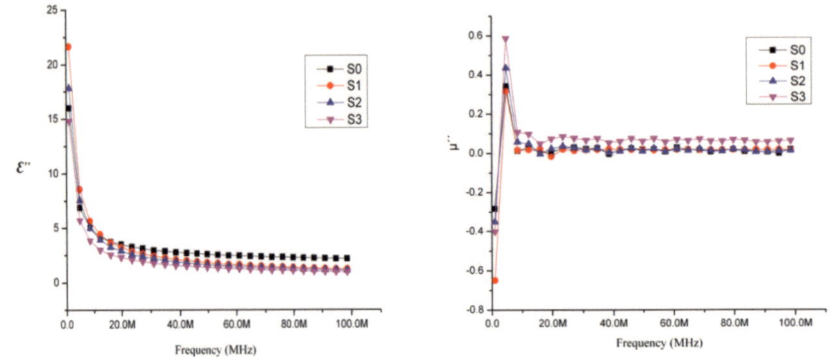

Figure 7.
Complex parts of permittivity and permeability of Co nanoferrite and composite samples in the frequency range of 1 MHz to 100 MHz.

low magnetic and dielectric graphene sheets [25]. Considering the spectra of complex permeability with respect to frequency as shown in **Figure 7**. We have some oscillating response from the zero mean point at lower MHz for all the composite samples which became stagnant at higher frequency. While cobalt ferrite has some rapid decrease and then became independent of higher frequency. The oscillatory response of composites is due to the presence of ferromagnetic nature of graphene sheet in cobalt ferrite matrix, at the start due to low incident frequency the magnetic field is not reflected hence negative values are recorded for loss of magnetic field by the samples. At higher frequency the ferrite response is dominated and goes up to some higher value then comes down and became linear in further frequency range. But again the overall response to magnetic loss is the same for all the four samples.

3.5 Microwave absorption

The reflectivity curves in the frequency range up to 3 GHz for RGO and cobalt ferrite samples are shown in **Figures 8** and **9**, with nearly equal thickness (1.9 ± 0.1 mm).

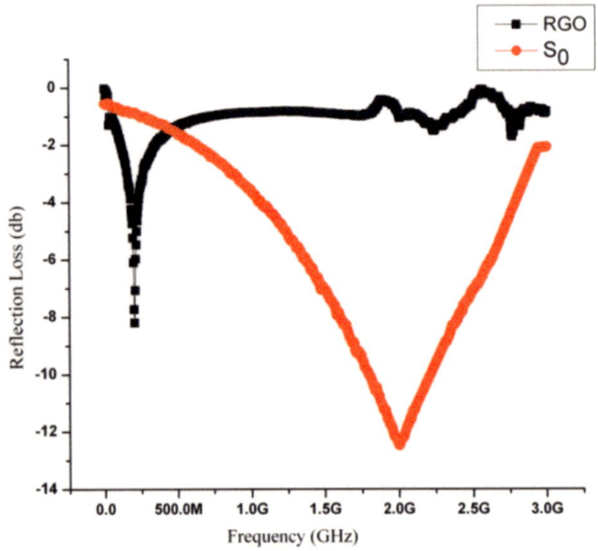

Figure 8.
Microwave absorption curves of cobalt and RGO samples with thickness 1.9 ± 0.1 mm.

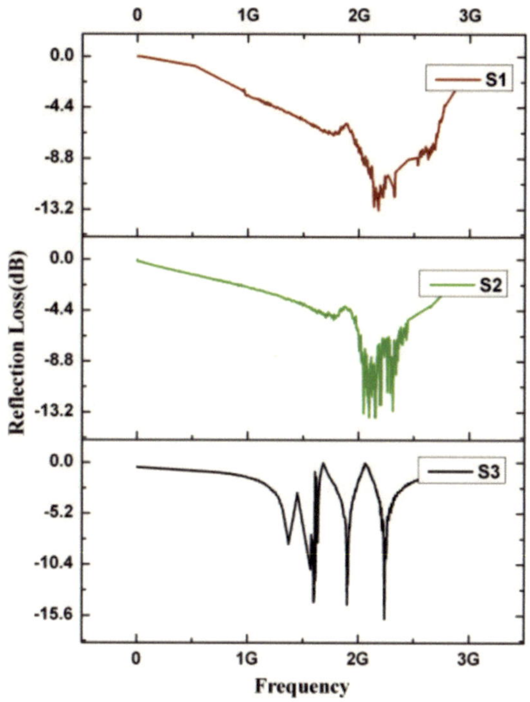

Figure 9.
Microwave absorption curves of Co composites samples with thickness 1.9 ± 0.1 mm.

Maximum absorption peak is reached at nearly 250 MHz at the value up to 9 db, while for ferrites we have greater absorption peak up to 13 db at very higher frequency (2GHz).Absorption spectra of composites samples are shown in **Figure 9**. It is clear that the position and value of absorption peak is not much displaced due to less percent doping of graphene (1%) but is very interesting that parent cobalt ferrite absorption peak is changed into a band of small peaks of nearly equal values.

Moreover as the doping of graphene is increased this band became widen. The reason behind is that single point maxima is converted to band of noise due to inhomogeneity and combine effects of electric (due to graphene), and magnetic dielectric nature (due to cobalt ferrites). So it is finally concluded that graphene based cobalt ferrites composites may be a potential candidate for microwave shielding.

4. Conclusion

In this study we have reported the cobalt ferrites nanocomposites embedded with graphene nanosheets, prepared by co-precipitation method. Various doping of graphene from 0.1% up to 1% were applied within the cobalt ferrite structure to study its microwave and mechanical effects on the nanocomposites. Microwave absorbing capacity of these samples was studied by Agilent network analyzer in low frequency band of microwave (1 MHz to 2 GHz), Results reveals that graphene incorporation not only improved the absorption capacity of cobalt ferrites(13 dB-17d), but also widened its maximum absorption peak. This change was supposed to be due to inhomogeneity and combine effects of electric (graphene), and magnetic dielectric nature (cobalt ferrites). We revealed that both the electrical properties and conductivity of cobalt ferrites is improved by imbedding very small amount of graphene nanosheet. Along with improving mechanical properties the graphene also widened the microwave absorption band. So, it is concluded that graphene based ferrites composites may play a vital role for microwave shielding applications.

Author details

Muhammad Siyar[1*] and Asghari Maqsood[2]

1 School of Chemical and Materials Engineering (SCME), National University of Science and Technology, NUST, Islamabad, Pakistan

2 Department of Physics, Nanoscale Physics Laboratory, Air University, Islamabad, Pakistan

*Address all correspondence to: muhammad.siyar@scme.nust.edu.pk

References

[1] V. Pallai and D.O Shah, Journal of Magnetism and Magnetic Materials 163, 243-248, (1996)

[2] R Skomski, J. Phys.: Condens. Matter 15, R1-R56, (2003).

[3] High Magnetostrictive Cobalt Ferrite for Sensor Application O. F. Caltun ,G.S.N.Rao ,K.H.Rao , B. Parvatheeswara Rao CheolGi Kim, Chong-Oh Kim, I. Dumitru , N. Lupu , and H. Chiriac sensor letters Vol.5, 1-3, 2007

[4] July 1999 Materials Letters 40 1999 39-45www.elsevier.com locate matlet

[5] G. Ranga Mohan a, D. Ravinder a , A.V. Ramana Reddy a , B.S. Boyanov ,Dielectric properties of polycrystalline mixed nickel–zinc ferrites 1999.

[6] l. Subramanian, m.n. Afsar, k.a. Korolev , Com 179 -20 millimeter wave material characterization of strontium ferrites using magneto-optical approach .

[7] A.S. Logginov, G.A. Meshkov, V.A. Nikolaev, A.P. Pyatakov , Magneto electric Control of Domain Walls in a Ferrite Garnet Film Physics Department, M.V. Lomonosov MSU, Moscow, Russia, 119992;

[8] W. Eerenstein, N. D. Mathur & J. F. Scott, Multiferroic and magnetoelectric materials, Nature, 442,759 (2006).

[9] Özgür et al., Microwave Ferrites, submitted to Journal of Materials Science: Materials in Electronics, 2009.

[10] E. Manova, D. Paneva, B. Kunev, Cl. Estournès, E. Rivière, K. Tenchev, A. Léaustic, I. Mitov. Mechanochemical synthesis and characterization of nanodimnsional iron-cobalt spinel oxides. Journal of Alloys and Compounds, 485 (1-2): 356-361, 2009

[11] Z. Zi, Y. Sun, X. Zhu, Z. Yang, J. Dai, W. Song. Synthesis and magnetic properties of CoFe O ferrite nanoparticles. Journal of Magnetism and Magnetic Materials, 321 (9): 1251-1255, 2009.

[12] H. Gul, A.Maqsood. Structural, magnetic and electrical properties of cobalt ferrites prepared by the sol-gel route. Journal of Alloys and Compounds, 465 (1-2): 227-231, 2008.

[13] K. Maaz, Arif Mumtaz, S.K. Hasanain, Abdullah Ceylan, Synthesis and magnetic properties of cobalt ferrite (CoFe2O4) nanoparticles prepared by wet chemical route, Journal of Magnetism and Magnetic Materials, Vol 308, 2, 289-295, ISSN 0304-8853, 2007.

[14] Q. Liu, J. Sun, H. Long, X. Sun, X. Zhong, Z. Xu. Hydrothermal synthesis of CoFe Onanoplatelets and nanoparticles. Materials Chemistry and Physics, 108 (2-3): 269-273, 2008.

[15] http://obsoletetellyemuseum.blogspot.com/2011/10/wega-color-3016-chassis-internal-view.html.

[16] I.Sharifi, H. Shokrollahi, M. M. Doroodmand, R. Safi, Magnetic and structural studies on CoFe2O4 nanoparticles synthesized by co-precipitation, normal micelles and reverse micelles methods, Journal of Magnetism and Magnetic Materials, 324 (10): 1854-1861,

[17] M Saidani a and MAM Gijs , applied physics letters volume numbe rmay 2004 High-quality radio-frequency inductor son silicon using a hybrid ferrite ,November 2003; 2004.

[18] W. Eerenstein, N. D. Mathur & J. F. Scott, Multiferroic and magnetoelectric materials, Nature, 442,759 (2006).

[19] Alain Tran, Miodrag Bolic, and Mustapha C.E. Yagoub , Magnetic-Field

coupling characteristics of ferrite-coil antennas for low-frequency RFID applications; IJCSI International Journal of Computer Scienc.7:2010.

[20] Zi, Z. Synthesis and magnetic properties of CoFe2O4 ferrite nanoparticles. J. Magn. Magn. Mater 2009; 321:1251-1255.

[21] Beny-Bassez and C Rouzaud J. N. Characterization of carbonaceous materials by correlated electron and optical microscopy and Raman micro spectroscopy. Scanning Electron Microscopy 1985; 1: 119-132.3.

[22] Rao, CNR, Sood, AK, Subrahmanyam and KS, Govindaraj. A: Graphene: the new two-dimensional nanomaterial. Angew Chem Int Ed 2009; 48: 7752-7777.

[23] J. C. Maxwell. A Dynamical Theory of the Electromagnetic Field. Philosophical Transactions of the Royal Society of London1865; 155: 459-512.

[24] B. V. Hamon. Maxwell-Wagner Loss and Absorption Currents in Dielectrics. Aust. J. Phys., Polarization1953; 6304.

[25] Peter Y. Yu, Manuel Cardona (2001). Fundamentals of Semiconductors: Physics and Materials Properties. Berlin: Springer. p. 261. ISBN 3-540-25470-6

Chapter 7

The Presented Study of Zn-Cu Ferrites for Their Application in "Photocatalytic Activities"

Sathyaseelan Balaraman, Baskaran Iruson, Senthilnathan Krishnmoorthy and Manikandan Elayaperumal

Abstract

The present chapter summarizes the synthesis and characterization of Zinc Copper (ZnCu) ferrites due to their wide range of applications in many areas. ZnCu ferrites are soft magnetic materials that have exceptional electrical, magnetic, and optical properties. ZnCu ferrites possess high resistivity, permeability, permittivity, saturation magnetization and low power losses, and coercivity. The above features of ZnCu ferrites find application in designing transformers, transducers, and inductors. Ferrites are also used in magnetic fluids, sensors, and biosensors. Apart from these advantages, they play a vital role in practical appliances like mobile, laptops, mobile chargers, refrigerators, washing machines, microwave ovens, printers, and so on. Therefore, the present focus is on the literature of techniques of the synthesis, their characterization, the effect of doping on the behavior of ZnCu ferrite, and, finally, their potential application in technology.

Keywords: spinel, ferrites, magnetic, surface morphology, synthesis techniques

1. Introduction

A ferrite is a ceramic material that is made up of iron oxide(Fe2O4) in massive proportion mixed with a metallic element such as barium (Ba), manganese (Mn), nickel (Ni), zinc (Zn) in small proportions [1–9]. The nature of iron oxide and the metal is electrically non-conducting and ferrimagnetic. Ferrimagnetic materials possess unequal opposing magnetic moments which help to retain spontaneous magnetization. Ferrites are generally classified into two types: hard ferrites [10, 11] and soft ferrites [12–15]. Hard ferrites have a high coercivity to magnetize. These materials are used in making permanent magnets which find application in the loudspeaker, washing machine, communication systems, switch-mode power supplies, dc-dc converters, microwave absorbing systems, and high-frequency applications [16–23]. On contrary, soft ferrites have low coercivity as a result of which their magnetization can easily be altered. Soft ferries are good conductors of the magnetic field which paved the way in developing transformer cores, high-frequency inductors, and microwave components [24–38]. Further, soft ferrites include high resistivity, low cost, temperature stability, low loss, and high permeability [39–42]. The most significant soft ferrites are ZnCu ferrites with high permeability and saturation magnetization [43–49]. Properties of these nanomaterials can be modified by substitution of diamagnetic or paramagnetic cations and their distribution in the spinel structure which alters the grain sizes, distribution, and surface morphology. The attribution of ZnCu ferrites is dependent on the synthesis methods and concentration of doping [50]. Cu-Zn spinel ferrites possess excellent magnetic behavior that depends on the concentration of Zn2+ ions in CuFe2O4. The large saturation magnetization (Ms), high initial permeability, and high resistivity of Cu-Zn ferrites make them a prominent material in electrical switches, memory devices, inductors, and antenna cores [51, 52]. Cu2+ has one unpaired electron, and Zn2+ has no unpaired electron. So, the replacement of Cu2+ ion by Zn2+ ion enhances the strength of magnetic ions participating in the exchange interaction. Moreover, the substitution of Zn2+ in place of Cu2 + changes the cation distribution and arrangement in the spinel structure. So, it is interesting to study the effect of Zn2+ substitution on the magnetic properties of Cu-ferrite nanoparticles. Moreover, Zn2+ substitution with higher concentration stabilizes the cubic phase of Cu-ferrite. The importance of Zn-Cu ferrites based on a comprehensive review is summarized.

2. Purpose of the review

The prime motive of the review is to focus on the synthesis and characterization methods of ZnCu ferrites. The subject of magnetic nanostructures is wide, and several articles are published. The ZnCu ferrites, constitute a special niche of nanoparticles because of the attraction of the scientific community towards soft ferrites. In addition, the study discusses the choice of techniques for synthesizing ZnCu ferrites by demand. The synthesis techniques such as the sol–gel [53], co-precipitation [54], a conventional ceramic technique [55–57], a hydrothermal method [58], citrate precursor [59], and microemulsion [60] for the synthesis of ZnCu ferrites are discussed.

3. Morphology of ZnCu ferrites

The spinel structure [61]. of ZnCu ferrite consists one major unit cell composed of 8 sub-unit cells having face-centered cubic (FCC) structure with two types of sites in each unit cell i.e. tetrahedral (A) site and octahedral (B) site. There are 64 tetrahedral interstitial sites and 32 octahedral interstitial sites. Spinel structure has closed packed oxygen atoms arrangement in which 32 oxygen atoms form a unit cell. Tetrahedral (A) sites are surrounded by four nearest oxygen atoms and octahedral (B) sites have six neighboring oxygen atoms. In ZnCu spinel ferrite lattice, Zn ions are on the tetrahedral sites while Fe and Cu ions occupy both tetrahedral and octahedral sites. Due to this spinel structure, different metallic ions can be added which brings out a change in the electric and magnetic properties of ferrites. The metal ions introduced may enter the spinel crystal lattice by replacingFe3+ ions and leading to aggregation of these ions on the grain boundary. These morphological features of ZnCu ferrite nanoparticles can be tuned to a larger extent to fulfill the application and to select the appropriate synthesis and characterization techniques. To learn the merits of ZnCu ferrites for various applications, the knowledge of different synthesis and characterization techniques is a must.

3.1 Synthesis methods to prepare ZnCu ferrites

There are two approaches to synthesize nanoparticles: top-down and bottom-up. In top-down, the bulk materials are broken down to get nanosized particles. This method has many shortcomings which include the presence of impurities, crystal defects, usage of metal oxides, the requirement of high temperature, products are inhomogeneous, broad size distribution, and flaws in surface structure. In the bottom-up approach, small atomic building blocks fuse to produce nanoparticles. This is the most conducive method for synthesis as the products are homogeneous, highly pure, and reveals narrow size distribution.

Various synthesis techniques are used to prepare ZnCu ferrite nanoparticles such as the sol–gel method [62], spray pyrolysis process [63], co-precipitation method [64], citrate precursor method [65]. By doping other elements or oxides the structural, electrical, and magnetic properties of ZnCu ferrite can be improved, leading to an increase in saturation magnetization and a decrease in coercivity in the soft ferrites.

3.1.1 Co-precipitation method

Co-precipitation [66, 67] is an easy and conventional method to synthesize nanomaterials. The ferrites prepared using this method are of controlled size, highly pure, and have a homogeneous structure. Normally inorganic salts (nitrate, chloride, sulfate, etc.) are used in this method as promoting substance that is dissolved in water or another medium that is suitable to form a homogeneous solution. The pH of the solution is adjusted to 7–9 and the solvent is evaporated to get nanoparticle precipitates. It should be observed that the concentration of salt, temperature, rate of change in pH are the parameters that determine the crystal growth and aggregation of the particles. After precipitation, the solid mass is collected and washed. By heating the residue the resultant product is dried to form hydroxides. The hydroxides are then transformed into crystalline oxides by calcination. Gautam et al. [68] used co-precipitation method to synthesize ZnCu ferrite. In this method, Copper chloride, zinc chloride, iron(III) chloride, and sodium hydroxide are used as raw materials. 3 M solution was prepared in 60 ml of distilled water which is mixed with boiling NaOH and stirred for 60 min between temperature 353–358 K using a magnetic stirrer, maintaining the pH between 11 and 12. The precipitates are allowed to settle down and then the sample was washed many times with distilled water. After washing, the sample was dried in hot air followed by crushing, and finally powdered with a mortar pestle A. C. Prabakar et.al [69] also synthesized ZnCu ferrites by the chemical co precipitation method by taking a solution of Cu(NO3)2.4H2O, Zn(NO3)2H2O, and Fe2(NO3)3.9H2O as the initial materials. These were mixed to form a homogeneous solution at 358 K. Then, ammonia solution was added dropwise with constant stirring maintaining the pH between 10 and 11. The mixture was heated at 353 K for 1 hr. After washing and drying the ferrite powder was heated at 673 K, 773 K, and 923 K separately and pressed in the form of circular pellets. Precipitates were washed many times and then dried in an oven at 100°C for 8 hr. The co-precipitation method has a unique advantage as it uses the aqueous medium for synthesis. There is an effective grain size control and morphology of the nanoparticles obtained.

3.1.2 Sol–gel method

The sol–gel method [70, 71] is a promising approach used for the preparation of ferrite nanomaterials. Sol is a colloidal suspension of solid particles of ions in a solvent and gel is a semi-rigid mass that forms when the solvent from the sol starts evaporating and the particles left behind join together. The resultant is in the form of colloidal powder or films. This method is appreciable due to the controlled microstructure. The final particles formed are of uniform and small size. Also, this technique of synthesis is economical and it is achievable at low temperature. The temperature conductivity was observed to be higher than pure ZnCu ferrite. There was a declination in the dielectric constant with an increase in nickel concentration. This method is advantageous because it is less time-consuming. The prepared ferrite consists of a pure cubic spinel structure.

3.1.3 Combustion method

The combustion process is an effective and low-cost method to synthesize nanomaterials. This process is simple, versatile, and fast for nanomaterial preparation. The salient feature of this method is, it requires less time and energy during the synthesis. The nanoparticles produced are pure and homogeneous. Researchers synthesized ZnCu ferrites by using this method [72, 73]. Copper nitrate [Cu(NO3)2.6H2O], zinc nitrate [Zn(NO3)2.6H2O], iron nitrate [Fe(NO3)3.9H2O] were taken in appropriate proportions and urea [CO(NH2)2] was used as a reducing agent in this process. A solution is formed by adding these materials in deionized water and heated at 480°C in air. Then, it is ignited within 5 s with a flame temperature of ~1600°C. The combustion technique is doping with other elements to decrease the lattice parameter, which reflects the fact that the ions of dopants get trapped at the grain limits. They hinder the development of grain and may cause an increase in strain on the grains that leads to a decrease in lattice parameters. The fuel chosen has a very vital impact on the ZnCu ferrites prepared. The fuels that are preferred in this method are urea and glycine, by which uniform nano ferrites with controlled stoichiometry are obtained.

3.2 Spectroscopic characterization

The characterizations of ZnCu ferrites are studied with an X-ray diffractometer, scanning electron microscopy [74–76]. The magnetic properties of the ferrites are studied by vibrating sample magnetometer (VSM), magnetization hysteresis (M – H) loops [77], and electron spin resonance (ESR) hysteresis loop measurements. The X-ray investigation is carried with an X-ray diffractometer with CuKα radiation (λ = 1.5405 Å). Many techniques are used to determine the shape, size, and morphology of magnetic nanoparticles such as XRD, SEM, TEM, HRTEM (High-resolution transmission electron microscopy), and FESEM (Field emission scanning electron microscopy). By using HRTEM we can get information about shape, size, crystallinity, and lattice spacing. XRD is used to determine the size by using the Scherrer equation. However, SEM is effective because the size of very small magnetic nanoparticles is determined.

3.3 XRD analysis

The crystalline structure of the Zinc copper ferrite as prepared and annealed at 500°C and 600°C are shown in **Figure 1**. All the diffraction peaks can be registered to the standard Zinc copper ferrite with the cubic spinel structure and the reflection peaks match adequately with standard data (JCPDS 82–1042). This indicates the synthesized nanoparticles have an Fd3m space group and indicating a high degree of purity. The average crystallite sizes, D, are calculated from the (311) peaks through Scherrer's formula around 32 nm. It is observed that the sample Zn1-x CuxFe2O4 crystallized in cubic structure and the lattice parameter is found to be 8.355 Å, which is less than the values reported for bulk compounds [78]. The crystallite size of ferrite nanoparticles is in the range of 2.3–11.8 nm and the average crystallite size and lattice constant were found to be 6.52 nm and 8.443 Å

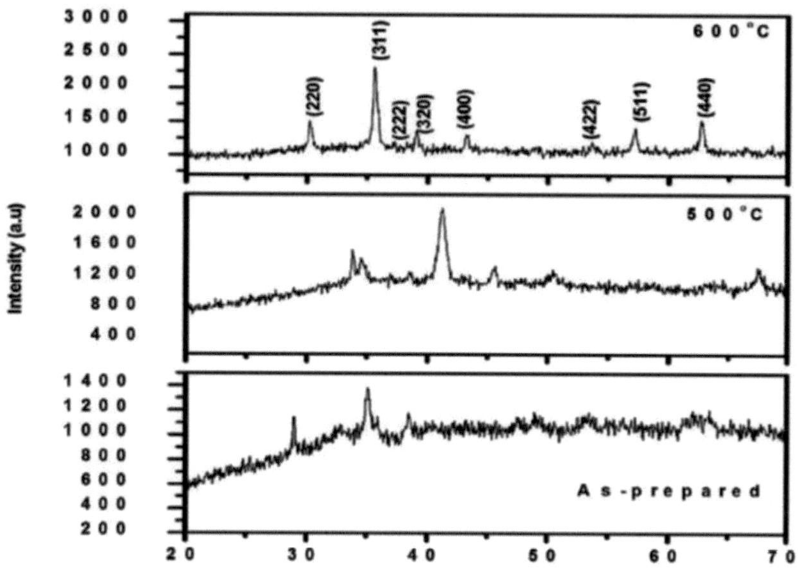

Figure 1.
XRD patterns of zinc copper ferrite nanostructures as prepared, annealed at 500°C, and 600°C.

respectively. Thomas Abo Atia et al. [79] studied the effect of sintering temperature and observed that the average crystallite size was found to increase with an increase in sintering temperature i.e. from 35 nm to 45 nm. Shwetambaram et al. [80]. Crystallite size was calculated and found to be of 7.74, 10.80, 11.58, 12 nm for x = 0.00, 0.25, 0.50, 0.75 respectively. Anuj Jain et al. [81] observed that the crystallite size of the composite ferrites increases on increasing the Cu concentration. This is due to the decrement in the densities of nucleation centers in the doped samples which result in the formation of larger crystallite size. S B Kale et al. [82]. The value of crystallite size was 24.17 nm indicates the nanocrystalline nature of the prepared samples. A. Subha et al. [83] observed from grain size calculated using Scherrer's formula applied to the most intense peak shows similar grain size was 30 nm. A. Tony Dhiwahara et al. [84] have indicated in the XRD patterns, the linear change in peak width was reflected in a linear decrease in crystallite size from 19.37 to 15.21 nm with an increase in Zn content in the host sublattice of $CuFe_2O_4$. This is mainly because of the replacement of smaller Cu^{2+} ions (0.72 Å) by larger Zn^{2+} ions (0.74 Å) [85, 86]. In addition, as the ionic radius of Zn^{2+} is larger than the ionic radius of Cu^{2+}, the Zn^{2+} substitution leads to a larger expansion of the lattice. Consequently, the lattice parameter increases more when compared to the Cu^{2+} substitution in the synthesized particles. Since ionic radius of Fe^{2+} (0.74 Å) ion is larger than Fe^{3+} ion (0.64 Å), the lattice constant increases [87]. Nonmagnetic transition metal ions Zn^{2+} and Cu^{2+} ion prefer octahedral sites whereas Fe^{3+} ions prefer both tetrahedral and octahedral sites. The characteristics peaks correlate with the ferrite particles and show the phase group Fd3m and spinel structure having a single phase. Hence, it is concluded that the ZnCu ferrites have a single-phase spinel cubic structure with an Fd3m phase group. However, some deviation in the structure can be observed because of doping.

3.4 Morphological structure

Various techniques such as TEM, and SEM, etc. are used to investigate the morphological structure of the ferrite nanoparticles. TEM is preferred because of its resolution. From the typical SEM analysis, morphological characteristics of Zinc copper ferrite nanoparticles annealed at 600°C are shown in **Figure 2(a)**. They show the formation of multigrain agglomerations consisting of fine crystallites with irregular shapes and sizes. Ferrite powders possessed a coarse structure with crystalline microstructure with an average grain size homogeneous is about 50 μm obtained from SEM images. This is larger than the size of nanocrystals calculated using the XRD measurements, which helps to understand that the grains are formed by aggregation of several nanocrystals. The samples are irregular shapes and sizes, and the cohesion of grains is due to the magnetic attraction. A drastic difference in microstructure of the annealed at 600°C products indicated that the substitution of metal ions like Zn, and Cu on the surface of this microstructure.

EDX analysis confirms the stoichiometric proportion of Zinc copper ferrite nanoparticles annealed at 600°C and also the percentage proportion of the constituent elements is shown in **Figure 2(b)**. The elemental weight proportion percentage is presented in the tables of weight and atomic percentage proportions, the constituent elemental proportion, and the ratios are in agreement with the expected elemental proportion and the oxygen (O) and iron (Fe) being with the highest peaks in all of the samples [88]. Typical EDX analysis reveals the existence of elements of Zn, Cu, Fe, and O. As portrayed in **Figure 3(a)**, HRTEM micrographs also confirm the particle size of ZnCu ferrite nanoparticles annealed at 600°C. The average crystallite size is around 8.38 nm. HRTEM analysis reveals that the particles are nearly spherical. The average crystallite size estimated from the HRTEM image falls in line with the observed values from powder XRD results. From **Figure 3(b)**, the HRTEM image of individual Zinc copper ferrite nanocrystal indicates that the interplanar distance is 0.26. In the SAED (**Figure 3(c)**) image of annealed at 600°C Zinc copper ferrite nanoparticles, the diffraction rings coincide with standards powder XRD diffraction data that confirms good crystallinity. The observed crystallographic d values of 2.52 Å correspond to the lattice space of the (311) plane of the Zinc copper ferrite system. The observed crystallographic d values agree well with those obtained from powder XRD analysis.

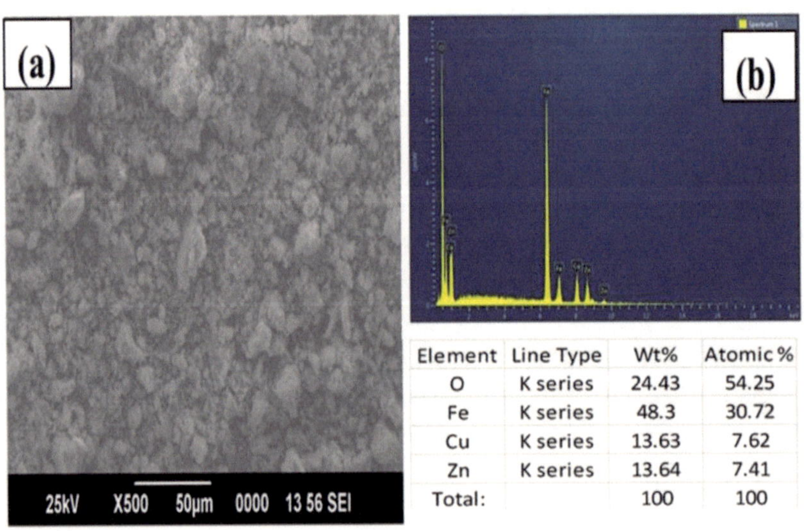

Element	Line Type	Wt%	Atomic %
O	K series	24.43	54.25
Fe	K series	48.3	30.72
Cu	K series	13.63	7.62
Zn	K series	13.64	7.41
Total:		100	100

Figure 2.
Images of zinc copper ferrite nanostructures annealed at 600°C (a) SEM (b) EDX analysis.

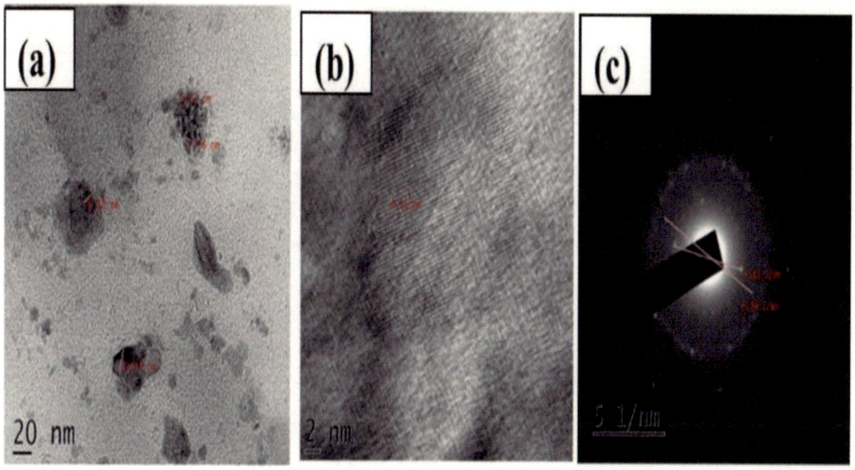

Figure 3.
Image of zinc copper ferrites nanostructures annealed at 600°C, (a) HRTEM (b) inter.

3.5 FT-IR analysis

FT-IR stands for Fourier transform Infrared, the method that is used for infrared spectroscopy. Some of the IR radiation is passed through a sample is absorbed by the sample and few are transmitted. The resulting spectrum represents the molecular absorption and transmission, creating a fingerprint of the sample. **Figure 4** shows the typical FT-IR spectra of Zinc copper ferrite nanostructures recorded between 4000 and 400 cm^{-1}. In the range of 1000–400 cm^{-1}, two main absorption bands of ferrite are appearing. The absorption band υ 1 - 580 cm^{-1} is assigned to the stretching vibration of tetrahedral complexes (Fe^{3+}-O^{2-}), and the absorption band υ2 = 400 cm^{-1} is attributed to the octahedral complexes (Fe^{3+}-O^{2-}). The peak intensity of υ 1 decreases with increasing Cu^{2+} contents, while the position band is shifted to lower frequencies. Synchronously, the intensity and position of υ 2 changed

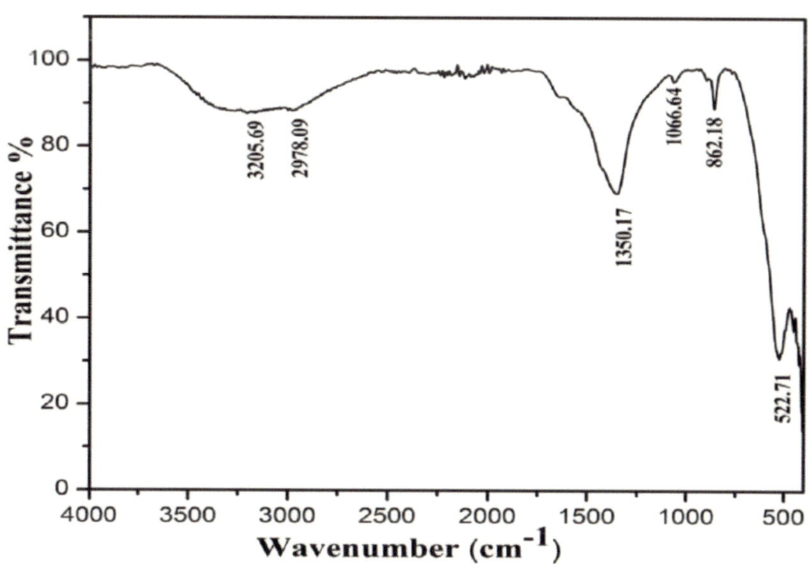

Figure 4.
FT-IR spectra of zinc copper ferrite nanostructures annealed sample at 600°C.

slightly with x. Similar results are observed in Zinc copper ferrite nanostructures prepared by the sol–gel combustion method [89, 90]. The difference in band position of υ1and υ2 can be related to the difference in Fe^{3+}-O^2 bond lengths at A sites and B sites. It was found that the Fe-O distance at A sites (1.89 Å) is smaller than that of the B sites (2.03 Å) [91, 92]. When Zn^{2+} ions are replaced by Cu^{2+} ions, due to charge imbalance some Fe^{3+} ions shift from A sites to B sites, making the Fe^{3+}-O^2 stretching vibration greater in the B domain. So the decrease in peak intensity of υ 1 with increasing Cu^{2+} content is mainly attributed to the change in Fe^{3+}-O^2 bands.

3.6 Magnetic properties

The most effective technique for analyzing the magnetic properties of ferrite nanoparticles are VSM (vibrating sample magnetometer), magnetization hysteresis (M – H) loop. One can find saturation magnetization, remanent magnetization, and coercivity by using these characterization techniques. The interior of the magnetic substance is normally divided into several domains, as the external magnetic field increases the domain walls may move and the magnetic field rotates within domains, resulting in a single-domain state. The magnetization saturation is attained if the magnetization axis and the external magnetic field direction are similar. The spontaneous magnetization (Ms) was obtained by extrapolating the high-field part of the loop to the zero applied field [93–97]. A similar variation of magnetization of ZnCu ferrite with the increase in Zn2+ concentration up to a particular x value has been reported in the literature [98–100]. Najmoddin et al. [101] observed the highest Ms. value of 52 emu/g at room temperature for x = 0.25 in ZnCu nanoferrites prepared by wet chemical method. Retentivity is the value of magnetization that is retained in the absence of an induced magnetic field. Coercivity is defined as the caliber of a ferromagnetic material to withstand an external magnetic field without demagnetizing it. In the case of a ferromagnetic material, it is defined as the intensity of the applied magnetic field that is required to reduce the magnetization to zero after the saturation state. The materials which have high coercivity are called hard materials and one with a low magnitude are soft materials. Hard materials are preferred for making permanent magnets whereas soft materials are used for making transformers, inductor cores, and microwave devices.

Figure 5 shows a typical hysteresis loop of all the Zinc copper ferrite nanostructures compositions which are annealed at 600°C. By using a vibrating sample magnetometer, the measurements of magnetization for all the compositions were carried out under the applied magnetic field of range ± 10 k Oe at room temperature. It is observed that the magnetic properties such as saturation magnetization (Ms), remanent magnetization (Mr), and coercivity (Hc) values are closer to the values inscribed in the literature. [102–106]. The structure of spinel ferrite is ferrimagnetic, the magnetic moments of A and B sites are coupled antiparallel to each other. Since there are doubly filled B sites, the net magnetic moment is the difference between the two sites. The magnetization behavior of spinel ferrite can be understood in Neel's model [107, 108]. The composition of Zinc copper ferrite nanostructures and cation distribution among the A and B sites will to a large degree, influence the magnetic properties of samples. According to this, in any ferrite, the magnetic order of tetrahedral clusters (A-site) and octahedral clusters (B-site) was found to be anti-parallel to each other. In this, the A-A and B-B superexchange interaction was predominated by the A-B superexchange interaction. According to Neel's model, the net magnetic moment of the composition per formula is expressed as:

$$\mu B = M_B(x) - M_A(x), \quad (1)$$

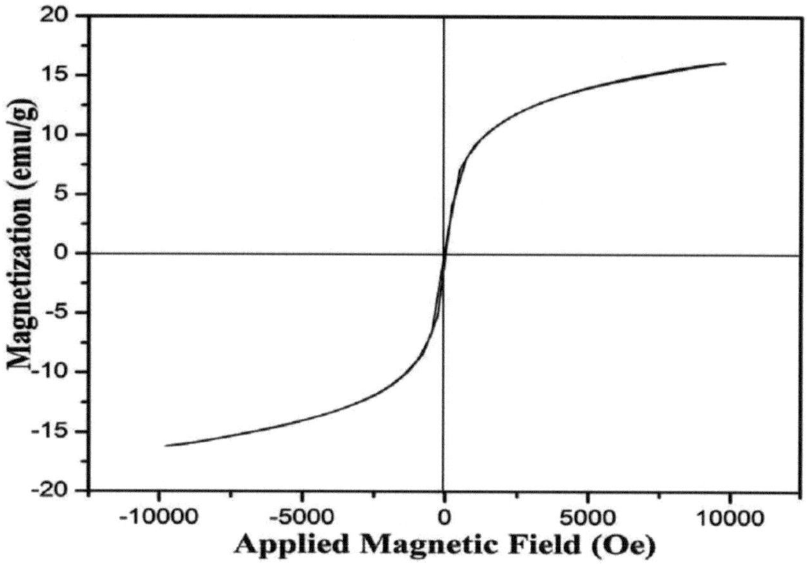

Figure 5.
Magnetic hysteresis loops for zinc copper ferrite nanostructures nanostructures annealed sample at 600 °C.

where M_B and M_A are the magnetic moments of B and A sublattices respectively.

The squareness ratio or remnant ratio (Mr/Ms) of a material is one of the important characteristics which depends on its anisotropy. The values of the squareness ratio represent the random arrangement of uniaxial particles along with the cubic magnetocrystalline anisotropy [109, 110]. In the study, the squareness ratio of pure ZnCuFe2O4 is 0.54, at room temperature. And it has been observed from the literature that the squareness, indicates the presence of non-interacting single domain particles with cubic anisotropy in the respective compositions [111].

The values of Bohr Magneton (μB) of these samples were also evaluated by using the following Equation.

$$\mu B = \frac{\text{MsMw}}{5585}$$

Where Mw is the molecular weight of the sample and Ms. is saturation magnetization.

5585 = β × N [β is Conversion factor (9.27 × 10–21); N is Avogadro's number].

It is observed that the value of Bohr Magneton is 1.56. One can adapt the composition of ferrite materials by squareness ratio (S) for the development of new electromagnetic materials and as per the need of the hour [112]. In the present work, the variation in the magnetic properties of Zinc copper ferrite nanostructures is obtained.

4. Photocatalytic degradation

The magnetic material ZnCuFe2O4 with the spinel structure has been proven effective in the application of dye removal. The visible light exposes, excellent photochemical stability. ZnCuFe2O4 has grabbed massive attention in the conversion of solar energy and photochemical hydrogen production from water. Also, the ZnCuFe2O4 magnetic particles possessed intrinsic peroxidase-like activity, which could react with H2O2 to produce •OH. Zinc Copper ferrite is one of the most important ferrites. ZnCuFe2O4 nanoparticles were found to be photo-sensitive in the visible light region (1.92 eV) with exceptional photochemical stability which paves way for them to act as gas sensors and photocatalysts [113, 114]. Photooxidation and photoreduction refer to the initiation of oxidation and reduction reactions by light energy. When irradiated with light energy, an electron (e-) is excited from the valence band (VB) to the conduction band (CB) of the photocatalyst, leaving a photogenerated hole (h+) photogenerated electron and holes are capable of oxidizing/reducing adsorbed substrates. The ZnCuFe2O4 NPs promote a photocatalytic reaction by serving as agents for the charge transfer between two adsorbed molecules. The charge transfer at the semiconductor–electrolyte interface is followed by bandgap excitation of a semiconductor nanoparticle. At the later stage, the nanoparticle quenches the excited state by accepting an electron, either transferring the charge to another substrate or generating photocurrent [115]. In both cases, the sensitivity of the semiconductor is retained which describes it as photocatalytic.

The photo-degradation phenomenon of methylene blue by Zinc copper ferrite nanocrystal is evaluated and the observed absorption spectra are presented in **Figure 6(a)**. It is clearly shown in the figure that the characteristic absorption

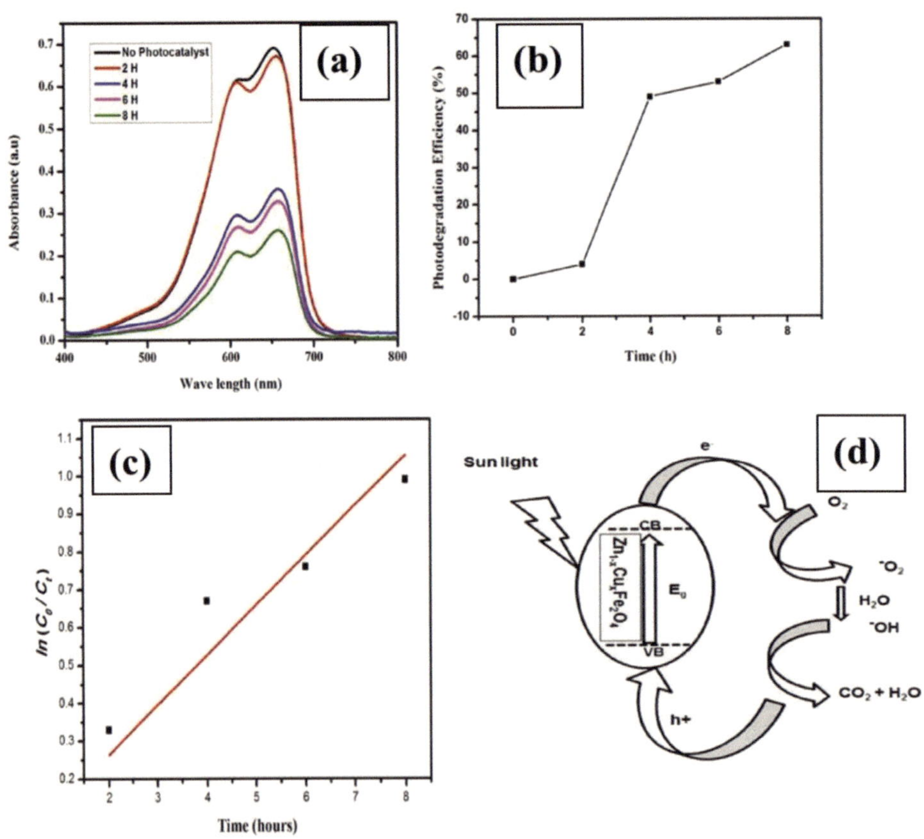

Figure 6.
Zinc copper ferrite annealed at 600°C under UV-light irradiation. (a) Absorption of MB solution during the photo-degradation (b) photo-degradation percentage (c) plots of ln[C/Co] versus irradiation time (d) photocatalytic mechanism of MB in the catalyst.

peak of the methylene blue (MB) at about 664 nm decay gradually with an enhanced exposure time of 8 hours and almost disappears after the irradiation time. This indicates that the MB dye has almost degraded. The photocatalytic performance of the NPs is observed by plotted (C/Co) as a function of time for MB dye and the same is shown in **Figure 6(b)**. The presence of Zinc copper ferrite nanocrystal emphasis the effective photo-degradation activity for MB, and no degradation of dye molecules was observed in the darkness. However, the Zinc copper ferrite sample exhibited 65% photodegradation. The variation of MB photo-degradation on the crystallite size of Zinc copper ferrite nanocrystals is shown in **Figure 6(c)**. Further, it reveals that the degradation percentage of MB and their kinetics [116, 117]. Photocatalytic reaction mechanisms for oxidation of MB dye by Zinc copper ferrite are presented in **Figure 6(d)**. Photo catalytic activities have been proved to advance by Zinc copper ferrite and it can be ascribed from the photo absorption spreading even up to the visible region, minimizing the electron–hole recombination rate.

4.1 ZnCu ferrites for ongoing COVID-19 pandemics

Nanomaterials are making a global impact on the health system and socio-economic progress. Nanoparticles of ZnCu have unique physical and chemical features that can be coupled with the development of potential therapeutic drugs, nanomaterial-based antiviral sprays, anti-viral surface coatings, and drug delivery. The study emphasizes the choice of synthesis method which decides the size and charges tunability properties to the ZnCu ferrites. The size tunability ensures that a large amount of drug can be fused into anatomically privileged sites of the virus, while charge tunability would facilitate the entry of the drug into charged parts of the virus. In addition, biosensors for the early detection of viral strains such as the COVID-19 can also be developed with ZnCu ferrites. For instance, ZnCu ferrites can be used to develop Giant magneto resistance-based sensors which have been applied in virus detection, earlier [114].

5. Outlook

The synthesis of ZnCu NPs among the soft ferrites with high saturation magnetization, low retentivity, low coercivity, high initial permeability, narrow size distribution, attracted scholars towards the fabrication of these ferrites. Co-precipitation is the most suitable strategy to obtain fine crystalline size among all the techniques. The XRD pattern of the ZnCu ferrites has characteristic peaks, reveals the cubic spinel phase having an Fd3m phase group. The shape of the prepared ferrite is nearly spherical but some distortion may be observed after doping. FTIR spectra confirmed the spinel phase of the ferrite nanoparticles having tetrahedral and octahedral sites. ZnCu ferrites have a lot of applications in the field of biomedicine, electronics, and for making ferrofluids, etc. To enhance the application of ZnCu ferrite nanoparticles, continuous and effective studies are required. The electrical and magnetic properties of ZnCu ferrites can be improved by doping with other functional metals such as cobalt, zinc, copper, and can be tested for agricultural and electrical applications. The nanoparticles during the pandemic outbreaks, like COVID-19, can play a significant role in the development of high contrast imaging dyes for viral strains in body fluids. ZnCuFe can also be preferred for developing nanomaterial-based medicines and therapeutics to serve the human community.

Author details

Sathyaseelan Balaraman[1*], Baskaran Iruson[2], Senthilnathan Krishnmoorthy[3] and Manikandan Elayaperumal[4,5,6]

1 Department of Physics, Thiruvalluvar University, Vellore, Tamil Nadu, India

2 Department of Physics, Arignar Anna Government Arts College, Cheyyar, Tamil Nadu, India

3 Department of Physics, VIT University, Vellore, Tamil Nadu, India

4 Department of Physics, Thiruvalluvar University, TVUCAS Campus, Thennangur, Tamil Nadu, India

5 Nanosciences African Network (NANO-AFNET), iThemba LABS-National Research Foundation, Somerset West, Western Cape Province, South Africa

6 UNESCO-UNISA Africa Chair in Nanoscience/Nanotechnology, College of Graduate Studies, University of South Africa, Pretoria, South Africa

*Address all correspondence to: bsseelan03@gmail.com

References

[1] Sugimoto, M.: The past, present, and future of ferrites. J. Am. Ceram. Soc. 82, 269-280 (1999). https://doi.org/10.1111/j.1551-2916.1999.tb20058.x.2. Bate, G.: Magnetic recording materials since 1975. J. Magn. Magn. Mater. 100, 413-424 (1991). https://doi.org/10.1016/0304-8853(91)90831-T3.

[2] Shylesh, S., Schünemann, V., Thiel, W.R.: Magnetically separablenanocatalysts: bridges between homogeneous and heterogeneouscatalysis. Angew. Chem. Int. Ed. 49, 3428-3459 (2010). https://doi.org/10.1002/anie.200905684.

[3] Veiseh, O., Gunn, J.W., Zhang, M.: Design and fabrication of magnetic nanoparticles for targeted drug delivery and imaging. Adv. Drug Deliv. Rev. 62, 284-304 (2010). https://doi.org/10.1016/j.addr.2009.11.002.5.

[4] Lee, J.-H., Huh, Y.-M., Jun, Y., Seo, J., Jang, J., Song, H.-T., Kim, S., Cho, E.-J., Yoon, H.-G., Suh, J.-S., Cheon, J.: Artificially engineered magnetic nanoparticles for ultra-sensitive molecular imaging. Nat. Med. 13, 95-99 (2007). https://doi.org/10.1038/nm1467.

[5] Lcc, J.H., Jang, J.T., Choi, J.S., Moon, S.H., Noh, S.H., Kim, J.W., Kim, J.G., Kim, I.S., Park, K.I., Cheon, J.: Exchange-coupled magnetic nanoparticles for efficient heat induction. Nat. Nanotechnol. 6,418-422 (2011). https://doi.org/10.1038/nnano.2011.95.7.

[6] Chen, R., Christiansen, M.G., Anikeeva, P.: Maximizing hysteretic losses in magnetic ferrite nanoparticles via model-driven synthesis and materials optimization. ACS Nano. 7, 8990-9000 (2013). https://doi.org/10.1021/nn4035266.

[7] Lu, A.H., Salabas, E.L., Schüth, F.: Magnetic nanoparticles: synthesis, protection, functionalization, and application. Angew. Chem. Int. Ed. 46, 1222-1244 (2007). https://doi.org/10.1002/anie.200602866.

[8] Dong, Y., He, K., Yin, L., Zhang, A.: A facile route to controlled synthesis of Co3O4 nanoparticles and their environmental catalytic properties. Nanotechnology. 18, 435602 (2007). https://doi.org/10.1088/0957-4484/18/43/435602.10.

[9] Masunga, N., Mmelesi, O.K., Kefeni, K.K., Mamba, B.B.: Recent advances in copper ferrite nanoparticles and nanocomposites synthesis, magnetic properties and application in water treatment: review. J. Environ. Chem. Eng. 7, 103179 (2019). https://doi.org/10.1016/j.jece.2019.103179.

[10] A. Namai, M. Yoshikiyo, K. Yamada, S. Sakurai, T. Goto, T. Yoshida, T. Miyazaki, M. Nakajima, T. Suemoto, H. Tokoro, S.I. Ohkoshi, Hard magnetic ferrite with a gigantic coercivity and high frequency millimetre wave rotation, Nat. Commun. 3 (2012) 1035-1036, https://doi.org/10.1038/ncomms2038.

[11] K.S. Martirosyan, N.S. Martirosyan, A.E. Chalykh, Structure and properties of hard-magnetic barium, strontium, and lead ferrites, Inorg. Mater. 39 (2003) 866-870, https://doi.org/10.1023/A:1025037716108.

[12] H. Anwar, A. Maqsood, Effect of sintering temperature on structural, electrical anddielectric parameters of Mn-Zn nano ferrites, Key Eng. Mater. 510-511 (2012) 163-170 https://doi.org/10.4028/www.scientific.net/kem.510-511.163.

[13] A.C.F.M. Costa, A.P.A. Diniz, A.G.B. de Melo, R.H.G.A. Kiminami, D.R. Cornejo, A.A. Costa, L. Gama,

Ni-Zn-Smnanopowder ferrites: morphological aspects and magnetic properties, J. Magn. Magn Mater. 320 (2008) 742-749, https://doi.org/10.1016/j.jmmm.2007.08.011.

[14] S.F. Mansour, O.M. Hemeda, M.A. Abdo, W.A. Nada, Improvement on the magnetic and dielectric behavior of hard/soft ferrite nanocomposites, J. Mol. Struct. 1152 (2018) 207-214, https://doi.org/10.1016/j.molstruc.2017.09.089.

[15] N. Modaresi, R. Afzalzadeh, B. Aslibeiki, P. Kameli, A.G. Varzaneh, I. Orue, Magnetic properties of Zn x Fe 3 − x O 4 nanoparticles : A competition between the effects of size and Zn doping level vol. 482, (2019), pp. 206-218.

[16] H.R. Kirchmayr, REVIEW ARTICLE Permanent magnets and hard magnetic materials, J. Appl. Phys. 29 (1996) 2763-2778.

[17] A. Chen, Y. Zhi, Y. Bao, Ceramic composites, J. Phys. Condens. Matter 6 (1999) 7921-7925, https://doi.org/10.1088/0953-8984/6/39/012

[18] S. Torkian, A. Ghasemi, R.S. Razavi, Magnetic properties of hard-soft SrFe10Al2O19/Co0.8Ni0.2Fe2O4 ferrite synthesized by one-pot sol-gel autocombustion, J. Magn. Magn Mater. 416 (2016) 408-416, https://doi.org/10.1016/j.jmmm.2016.05.050.

[19] F.J. Heiligtag, M. Niederberger, The fascinating world of nanoparticle research, Mater. Today 16 (2013) 262-271, https://doi.org/10.1016/j.mattod.2013.07.004.

[20] H. Stäblein, Hard ferrites and plastoferrites, Handb. Ferromagn. Mater. 3 (1982) 441-602, https://doi.org/10.1016/S1574-9304(05)80093-8.

[21] T.S. Chin, Permanent magnet films for applications in microelectromechanical systems, J. Magn. Magn Mater. 209 (2000) 75-79, https://doi.org/10.1016/S0304-8853(99)00649-6.

[22] S. He, H. Zhang, Y. Liu, F. Sun, X. Yu, X. Li, L. Zhang, L. Wang, K. Mao, G. Wang, Y. Lin, Z. Han, R. Sabirianov, H. Zeng, Maximizing specific loss power for magnetic hyperthermia by hard–soft mixed ferrites, Small 14 (2018) 1-9, https://doi.org/10.1002/smll.201800135.

[23] Haining Ji, Zhongwen Lan, Yu Zhong, Zhiyong Xu, Microstructure and temperature dependence of magnetic properties of MnZnTiSn ferrites for power applications, IEEE Trans. Magn. 46 (2009) 974-978, https://doi.org/10.1109/tmag.2009.2037954.

[24] C.I. Covaliu, D. Berger, C. Matei, L. Diamandescu, E. Vasile, C. Cristea, V. Ionita, H. Iovu, Magnetic nanoparticles coated with polysaccharide polymers for potential biomedical applications, J. Nanoparticle Res. 13 (2011) 6169-6180, https://doi.org/10.1007/s11051-011-0452-6.

[25] M.J. Carey, S. Maat, P. Rice, R.F.C. Farrow, R.F. Marks, A. Kellock, P. Nguyen, B.A. Gurney, Spin valves using insulating cobalt ferrite exchange-spring pinning layers, Appl. Phys. Lett. 81 (2002) 1044-1046, https://doi.org/10.1063/1.1494859.

[26] W.A. Kaczmarek, B.W. Ninham, Application of mechanochemistry in ferrite materials technology, J. Phys. IV (1997), https://doi.org/10.1051/jp4:1997106 C1-47-C1-48.

[27] A. Zapata, G. Herrera, Effect of zinc concentration on the microstructure and relaxation frequency of Mn-Zn ferrites synthesized by solid state reaction, Ceram. Int. 39 (2013) 7853-7860, https://doi.org/10.1016/j.ceramint.2013.03.046.

[28] G. Herrera, Domain wall dispersions: relaxation and resonance in

Ni-Zn ferrite doped with V2 O3, J. Appl. Phys. 108 (2010) 3-8, https://doi.org/10.1063/1.3506716.

[29] S.C.Ó. Mathúna, T. O'Donnell, N. Wang, K. Rinne, Magnetics on silicon: an enabling technology for power supply on chip, IEEE Trans. Power Electron. 20(2005) 585-592, https://doi.org/10.1109/TPEL.2005.846537.

[30] P. Mathur, A. Thakur, M. Singh, P HYSICAL J OURNAL nano-ferrite and characterization for high frequency 138 (2008) 133-138, https://doi.org/10.1051/epjap.

[31] A. Hajalilou, S.A. Mazlan, A review on preparation techniques for synthesis of nanocrystalline soft magnetic ferrites and investigation on the effects of microstructure features on magnetic properties, Appl. Phys. Mater. Sci. Process 122 (2016), https://doi.org/10.1007/s00339-016-0217-2.

[32] V. Marghussian, Magnetic Properties of Nano-Glass Ceramics (2015), https://doi.org/10.1016/b978-0-323-35386-1.00004-9.

[33] V.S. Coker, N.D. Telling, G. Van Der Laan, R.A.D. Pattrick, C.I. Pearce, E. Arenholz, F. Tuna, R.E.P. Winpenny, J.R. Lloyd, Harnessing the extracellular bacterial production of nanoscale cobalt ferrite with exploitable magnetic properties, ACS Nano 3 (2009) 1922-1928, https://doi.org/10.1021/nn900293d.

[34] P.M. Tamhankar, A.M. Kulkarni, S.C. Watawe, Functionalization of cobalt ferrite nanoparticles with alginate coating for biocompatible applications, Mater. Sci.Appl. (2011) 1317-1321, https://doi.org/10.4236/msa.2011.29179 02.

[35] H. Zhao, C. Ragusa, C. Appino, O. De La Barriere, Y. Wang, F. Fiorillo, Energy losses in soft magnetic materials under symmetric and asymmetric induction waveforms, IEEE Trans. Power Electron. 34 (2019) 2655-2665, https://doi.org/10.1109/TPEL.2018.2837657.

[36] J. Füzer, M. Strečková, S. Dobák, Ďáková, P. Kollár, M. Fáberová, R. Bureš, Y. Osadchuk, P. Kurek, M. Vojtko, Innovative ferrite nanofibres reinforced soft magnetic composite with enhanced electrical resistivity, J. Alloys Compd. 753 (2018) 219-227, https://doi.org/10.1016/j.jallcom.2018.04.237.

[37] R. Justin Joseyphus, A. Narayanasamy, K. Shinoda, B. Jeyadevan, K. Tohji, Synthesis and magnetic properties of the size-controlled Mn-Zn ferrite nanoparticles by oxidation method, J. Phys. Chem. Solid. 67 (2006) 1510-1517, https://doi.org/10.1016/j.jpcs.2005.11.015.

[38] A.Z. Simões, C.S. Riccardi, M.L. Dos Santos, F.G. Garcia, E. Longo, J.A. Varela, Effect of annealing atmosphere on phase formation and electrical characteristics of bismuth ferrite thin films, Mater. Res. Bull. 44 (2009) 1747-1752, https://doi.org/10.1016/j.materresbull.2009.03.011.

[39] M. Pardavi-Horvath, Microwave applications of soft ferrites, J. Magn. Magn Mater.(2000), https://doi.org/10.1016/S0304-8853(00)00106-2.

[40] J. Töpfer, J. Mürbe, A. Angermann, S. Kracunovska, S. Barth, F. Bechtold, Soft ferrite materials for multilayer inductors, Int. J. Appl. Ceram. Technol. 3 (2006) 455-462, https://doi.org/10.1111/j.1744-7402.2006.02110.x.

[41] C. Miclea, C. Tanasoiu, C.F. Miclea, A. Gheorghiu, V. Tanasoiu, Soft ferrite materials for magnetic temperature transducers and applications, J. Magn. Magn Mater. 290-291 (2005) 1506-1509, https://doi.org/10.1016/j.jmmm.2004.11.561.

[42] S. Son, R. Swaminathan, M.E. McHenry, Structure and magnetic

properties of rf thermally plasma synthesized Mn and Mn–Zn ferrite nanoparticles, J. Appl. Phys.93 (2003) 7495-7497, https://doi.org/10.1063/1.1557953.

[43] Surendra Singh, Navendu Goswami, and S. C. Katyal, Nanostructured Zn-Cu ferrite: Structural, magnetic properties and application, AIP Conference Proceedings 2009, 020022 (2018); doi: 10.1063/1.5052091.

[44] Al-Gaashani, R.; Aïssa, B.; Anower Hossain, M.; Radiman, S. Catalyst-Free Synthesis of ZnO-CuO-ZnFe2O4 Nanocomposites by a Rapid One-Step Thermal Decomposition Approach. Mater. Sci. Semicond. Process. 2019, 90, 41–49.

[45] Lv, W.; Liu, B.; Luo, Z.; Ren, X.; Zhang, P. XRD Studies on the Nanosized Copper Ferrite Powders Synthesized by Sonochemical Method. J. Alloys Compd. 2008, 465, 261–264.

[46] Kurian, J.; Mathew, M. J. Structural, Optical and Magnetic Studies of CuFe2O4, MgFe2O4 and ZnFe2O4 Nanoparticles Prepared by Hydrothermal/Solvothermal Method. J. Magn. Magn. Mater. 2018, 451, 121–130.

[47] Zakiyah, L. B.; Saion, E.; Al-Hada, N. M.; Gharibshahi, E.; Salem, A.; Soltani, N.; Gene, S. Up-Scalable Synthesis of Size-Controlled Copper Ferrite Nanocrystals by Thermal Treatment Method. Mater. Sci. Semicond. Process. 2015, 40, 564–569.

[48] López-Ramón, M. V.; Álvarez, M. A.; Moreno-Castilla, C.; Fontecha-CÁmara, M. A.; Yebra-Rodríguez, Á.; Bailón-García, E. Effect of Calcination Temperature of a Copper Ferrite Synthesized by a Sol-Gel Method on Its Structural Characteristics and Performance as Fenton Catalyst to Remove Gallic Acid from Water. J. Colloid Interface Sci. 2018, 511, 193–202.

[49] Zhuravlev, V. A.; Minin, R. V.; Itin, V. I.; Lilenko, I. Y. Structural Parameters and Magnetic Properties of Copper Ferrite Nanopowders Obtained by the Sol-Gel Combustion. J. Alloys Compd. 2017, 692, 705–712.

[50] Tatarchuk, T., Paliychuk, N., Pacia, M., Kaspera, W., Macyk, W., Kotarba, A., Bogacz, B.F., Pędziwiatr, A.T., Mironyuk, I., Gargula, R., Kurzydło, P., Shyichuk, A.: Structure–redox reactivity relationships in Co1−xZnxFe2O4: the role of stoichiometry. New J. Chem. 43, 3038-3049 (2019).

[51] M. Rana, M. Islam, T. Abbas, Cation distribution and magnetic interactions in Zn-substituted CuFe2O4 ferrites, Mater. Chem. Phys. 65 (3) (2000) 345-349.

[52] M. Ajmal, A. Maqsood, Structural, electrical and magnetic properties of Cu1−xZnxFe2O4 ferrites ($0 \leq x \leq 1$), J. Alloy. Compd. 460(1-2) (2008) 54-59.

[53] Isfahani, M.J.N et al., Magnetic behavior of nickel-bismuth ferrite synthesized by a combined sol-gel/thermal method, Ceramics International, 39, (2013), pp. 1163-1167.

[54] Chen, D.G., Tang, X.G., Wu, J.B., Zhang, W., Liu, Q.X., Jiang, Y.P.: Effect of grain size on the magnetic properties of superparamagnetic Ni0.5Zn0.5 Fe2O4 nanoparticles by coprecipitation process. J. Magn. Magn. Mater. 323, 1717-1721(2011).

[55] Kulkarni, R.G., Patil, V.U.: Magnetic ordering in Cu-Zn ferrite. J.Mater. Sci. 17, 843-848 (1982). https://doi.org/10.1007/BF00540382.

[56] Rana, M.U., Islam, M., Abbas, T.: Cation distribution and magneticinteractions in Zn-substituted CuFe2O4 ferrites. Mater. Chem.Phys. 65, 345-349 (2000). https://doi.org/10.1016/S0254-0584(00)00218-2.

[57] Maria, K.H., Choudhury, S., Hakim, M.A.: Structural phase transformation and hysteresis behavior of Cu-Zn ferrites. Int. Nano Lett.3, 42 (2013). https://doi.org/10.1186/2228-5326-3-42.

[58] Shahida, A., Paul, D.P., Hoque, S.M., Hakim, M.A., Hudl, M.,Mathieu, R., Nordblad, P.: Magnetic and magnetocaloric properties of Cu 1_xZnxFe2O4 (x=0.6, 0.7,0.8) ferrites. J. Magn. Magn. Mater. 367, 75-80 (2014).

[59] Tatarchuk, T., Bououdina, M., Vijaya, J.J., Kennedy, L.J.: SpinelFerrite Nanoparticles: Synthesis, Crystal Structure, Properties, and Perspective Applications. Springer Proceedings in Physics. 195,305-325 (2017).

[60] Bogacz, B.F., Gargula, R., Kurzydło, P., Pędziwiatr, A.T., Tatarchuk, T., Paliychuk, N.: Two-level model description of superparamagnetic relaxation in nanoferrites (Co,Zn) Fe2O4. Acta Phys. Pol. A. 134(2018), 993-997 (2018).

[61] W. Shen, L. Zhang, B. Zhao, Y. Du, X. Zhou, Growth mechanism of octahedral like nickel ferrite crystals prepared by modified hydrothermal method and morphology dependent magnetic performance, Ceram. Int. 44 (2018) 9809-9815. https://doi.org/10.1016/j.ceramint.2018.02.219.

[62] J. Venturini, R.Y. Sun-Zampiva, S. Arcaro, C. Perez-Bergmann, Sol-gel synthesis of substoichiometric cobalt ferrite (CoFe2O4) spinels: influence of additives on their toichiometry and magnetic properties, Ceram. Int. 11 (2018) 12381-12388. https://doi.org/10.1016/j.ceramint.2018.04.026.

[63] D. Kotsikau, M. Ivanovskaya, V. Pankov, Y. Fedotova, Structure and magnetic properties of manganese–zinc-ferrites prepared by spray pyrolysis method, Solid State Sci. 39 (2019) 69-73. https://doi.org/10.1016/j.solidstatesciences.2014.11.013.

[64] A. Yusmar, L. Armitasari, E. Suharyadi, Effect of Zn on dielectric properties of Mn-Zn spinel ferrite synthesized by coprecipitation, Mater. Today Proc. 5 (2018) 14955-14959. https://doi.org/10.1016/j.matpr.2018.04.037.

[65] Tatarchuk, T., Paliychuk, N., Pacia, M., Kaspera, W., Macyk, W., Kotarba, A., Bogacz, B.F., Pędziwiatr, A.T.,Mironyuk, I., Gargula, R., Kurzydło, P., Shyichuk, A.: Structure–redox reactivity relationships in Co1–xZnxFe2O4: the role of stoichiometry. New J. Chem. 43, 3038-3049 (2019).

[66] Verma, K., Kumar, A., Varshney, D.: Effect of Zn and Mg doping on structural, dielectric and magnetic properties of tetragonal CuFe 2 O 4. Curr. Appl. Phys. 13, 467-473 (2013). https://doi.org/10.1016/j.cap.2012.09.015.

[67] Hankare, P.P., Kadam, M.R., Patil, R.P., Garadkar, K.M., Sasikala, R., Tripathi, A.K.: Effect of zinc substitution on structural and magnetic properties of copper ferrite. J. Alloys Compd. 501, 37-41(2010). https://doi.org/10.1016/j.jallcom.2010.03.178.

[68] Gautam, N., Thirupathi, G., Singh, R.: Effcct of Zn-doping onstructural and magnetic properties of copper ferrite nanoparticles. AIP Conf. Proc. 1731, (2016). https://doi.org/10.1063/1.4947753.

[69] A.C. Prabakar, B. Sathyaseelan ,G. Killivalavan , I. Baskaran , K. Senthilnathan, E. Manikandan, D. Sivakumar, Photocatalytic Dye Degradation Properties of Zinc Copper Ferrites Nanoparticles, Nanostruct, 2019; 9(4): 694-701. DOI: 10.22052/JNS.2019.04.011.

[70] Atif, M., Nadeem, M., Grössinger, R., Turtelli, R.S.: Studies on the magnetic, magnetostrictive and electrical properties of sol-gel

synthesized Zn doped nickel ferrite. J. Alloys Compd. 509, 5720-5724 (2011). https://doi.org/10.1016/j.jallcom.2011.02.163.

[71] Subha, A., Shalini, M.G., Sahu, B., Sahoo, S.C.: Structural transformation and magnetic properties of copper ferrite nanoparticles prepared by sol–gel method. J. Mater. Sci. Mater. Electron. 29,20790-20799 (2018). https://doi.org/10.1007/s10854-018-0221-8.

[72] Mangalaraja, R.V.; Anathakmar, S.; Manohar, P.; Gnanam, F.D.; Awana, M. Characterization of Mn0.8Zn0.2Fe2O4synthesized by flash combustion technique. Mater. Sci. Eng. A 2004, 367, 301-305.

[73] Kanagesan, S.; Hashim, M.; Tamilselvan, S.; Alitheen, N.B.; Ismail, I.; Syazwan, M.; Zuikimi I, M.M.M. Sol-gel auto-combustion synthesis of cobalt ferrite and it's cytotoxicity properties. Digest J. Nanomater. Biostruct. 2013, 8, 1601-1610.

[74] V. Loyau, G.Y. Wang, M. Lo Bue, F. Mazaleyrat, An analysis of MnZn ferrite microstructure by impedance spectroscopy, scanning transmission electron microscopy and energy dispersion spectrometry characterizations, J. Appl. Phys. 111(2012), https://doi.org/10.1063/1.3693544.

[75] A.H. Lu, E.L. Salabas, F. Schüth, Magnetic nanoparticles: synthesis, protection, functionalization, and application, Angew. Chem. Int. Ed. 46 (2007) 1222-1244, https://doi.org/10.1002/anie.200602866.

[76] D. Kotsikau, M. Ivanovskaya, V. Pankov, Y. Fedotova, Structure and magnetic properties of manganese-zinc-ferrites prepared by spray pyrolysis method, Solid State Sci. 39 (2015) 69-73, https://doi.org/10.1016/j.solidstatesciences.2014.11.013.

[77] R. Szewczyk, Model of the magnetostrictive hysteresis loop with local maximum, Materials 12 (2018) 105, https://doi.org/10.3390/ma12010105.

[78] Surendra Singh, Navendu Goswami, and S. C. Katyal, Nanostructured Zn-Cu Ferrite: Structural, Magnetic Properties and Application, AIP Conference Proceedings 2009, 020022 (2018); doi: 10.1063/1.5052091.

[79] Abo Atia T., Altimari P., Moscardini E., Pettiti I. , Toro L., Pagnanelli F., 2016, Synthesis and characterization of copper ferrite magnetic nanoparticles by hydrothermal route, Chemical Engineering Transactions, 47, 151-156 DOI: 10.3303/CET1647026.

[80] Shwetambaram, Anuj Jain, Z. Vakil, Synthesis of Zinc-Copper Nano-Ferrite and Characterization of their Structural, Electrical and Magnetic Behaviors, Int. J. Chem Tech Res. 2014, 6(3), pp 2207-2209.

[81] Anuj Jain, Ravi Kant Baranwa, Ajaya Bharti, Z. Vakil, and C. S. Prajapati, Study of Zn-Cu Ferrite Nanoparticles for LPG Sensing, The Scientific World Journal Volume 2013, Article ID 790359, 7 pages http://dx.doi.org/10.1155/2013/790359.

[82] S B Kale, R M Borade, J S Kounsalye A V Raut, S R Nimbhoreand K M Jadhav, Structural and dielectric properties of mixed spinel ferrite Cu(0.7)Zn(0.3) Fe2O4 nanoparticles, Journal of Physics: Conference Series 1644 (2020) 012012,doi:10.1088/1742-6596/1644/1/012012.

[83] A. Subha, M. Govindaraj Shalini1, Shantinarayan Rout, Subasa C. Sahoo, Magnetic Ordering and Enhancement of Magnetizationin Zinc-Substituted Copper Ferrite Nanoparticles, Journal of Superconductivity and Novel Magnetism https://doi.org/10.1007/s10948-020-05613-z.

[84] A. Tony Dhiwahara, , M. Sundararajanb, P. Sakthivel, Chandra Sekhar Dash, S. Yuvaraj, Microwave-assisted combustion synthesis of pure and zinc-doped copper ferrite nanoparticles: structural, morphological, optical, vibrational, and magnetic behavior, Journal of Physics and Chemistry of Solids (2019), doi: https://doi.org/10.1016/j.jpcs.2019.109257.

[85] Tan X, Li G, Zhao Y, Hu C. Effect of Cu content on the structure of Ni1-x CuxFe$_2$O$_4$spinels. Mater. Res. Bull, 2009; 44: 2160-2168.

[86] Zhang H. E, Zhang B. F, Wang G. F, Dong X. H, Gao Y. The structure and magnetic properties of Ni1-xZnxFe$_2$O$_4$ ferrite nanoparticles prepared by sol-gel auto-combustion. J. Magn. Magn. Mater, 2007; 312: 126-130.

[87] Denton A. R, Ashcroft N.W. Vegards law. Phy. Rev. A, 1991; 43: 3161-3164.

[88] Kim W.C, Kim S.J, Lee S.W, Kim C.S. Growth of ultrafine NiZnCu ferrite and magnetic properties by a sol–gel method. J. Magn. Magn Mater, 2001; 226: 1418-1420.

[89] Tan, X.Y., Li, G.Y., Zhao, Y. and Hu, C.W. The Effect of Cu Content on the Structure of Ni1−xCuxFe2O4 Spinels. Materials Research Bulletin, 2009;44:2160-2168.

[90] Xiang, J., Shen, X.Q., Song, F.Z. and Liu, M.Q. One-Dimensional NiCuZnFerrite Nanostructures: Fabrication, Structure, and Magnetic Properties. Journal of Solid State Chemistry, 2010;183: 1239-1244.

[91] Umare SS, Ningthoujam RS, Sharma SJ, Shrivastava S, Kurian S, Gajbhiye NS. Mössbauer and magnetic studiesonnanocrystalline NiFe2O4 particles prepared by ethyleneglycol route. Hyperfine Interactions. 2008;184(1-3):235-243.

[92] Rais A, Taibi K, Addou A, Zanoun A, Al-Douri Y. Copper substitution effect on the structural properties of nickelferrites. Ceramics International. 2014;40(9):14413-9.

[93] Chithra, M., Anumol, C.N., Argish, V., Sahu, B., Sahoo, S.C.:Tailoring magnetic properties of cobalt ferrite nanoparticles by differentdivalentcation substitution. J. Mater. Sci. Mater. Electron. 29, 813-822 (2018). https://doi.org/10.1007/s10854-017-7976-1.34.

[94] Chithra, M., Anumol, C.N., Sahu, B., Sahoo, S.C.: Structural and magnetic properties of Zn X Co 1−X Fe 2 O 4 nanoparticles: nonsaturation of magnetization. J. Magn. Magn. Mater. 424, 174-184 (2017). https://doi.org/10.1016/j.jmmm.2016.10.064.35.

[95] Dash, J., Prasad, S., Venkataramani, N., Krishnan, R., Kishan, P., Kumar, N., Kulkarni, S.D., Date, S.K.: Study of magnetization and crystallization in sputter deposited LiZn ferrite thin films. J. Appl. Phys. 86, 3303-3311 (1999). https://doi.org/10.1063/1.371206.36.

[96] Kulkarni, P.D., Prasad, S., Venkataramani, N., Krishnan, R., Pang, W., Guha, A., Woodward, R.C., Stamps, R.L.: The high field magnetization in the RF sputter deposited copper ferrite thin films. Proc.9th Int. Conf. Ferrites, p. 201. [25]. Wiley, San Fr. 2004 (2010).

[97] Sahoo, S.C., Venkataramani, N., Prasad, S., Bohra, M., Krishnan, R.: Stability of nonthermodynamic equilibrium cation distribution frozen during pulsed laser deposition of Co-ferrite thin films. Appl. Phys. A Mater. Sci. Process. 98, 889-894 (2010). https://doi.org/10.1007/s00339-009-5471-0.

[98] Kulkarni, R.G., Patil, V.U.: Magnetic ordering in Cu-Zn ferrite. J.Mater. Sci. 17, 843-848 (1982). https://doi.org/10.1007/BF00540382,

[99] Maria, K.H., Choudhury, S., Hakim, M.A.: Structural phase transformationand hysteresis behavior of Cu-Zn ferrites. Int. Nano Lett.3, 42 (2013). https://doi.org/10.1186/2228-5326-3-42,

[100] Hankare, P.P., Kadam, M.R., Patil, R.P., Garadkar, K.M., Sasikala, R., Tripathi, A.K.: Effect of zinc substitution on structural and magnetic properties of copper ferrite. J. Alloys Compd. 501, 37-41 (2010). https://doi.org/10.1016/j.jallcom.2010.03.178.

[101] Najmoddin, N., Beitollahi, A., Devlin, E., Kavas, H., Mohseni, S.M., Åkerman, J., Niarchos, D., Rezaie, H., Muhammed, M., Toprak, M.S.: Magnetic properties of crystalline mesoporous Znsubstituted copper ferrite synthesized under nanoconfinement in silica matrix. Microporous Mesoporous Mater. 190, 346-355 (2014). https://doi.org/10.1016/j.micromeso.2014.02.033.

[102] Maaz. K. and Mumtaz. A. Synthesis and Magnetic Properties of Cobalt Ferrite (CoFe2O4) Nanoparticles Prepared by Wet Chemical Route. Journal of Magnetism and Magnetic Materials, 2007;308:289-295.

[103] C. Sasikala, G. Suresh, N. Durairaj, I. Baskaran, B. Sathyaseelan, M. Kumar, K. Senthilnathan, E. Manikandan, Influences of Ti4+ ion on dielectric property in perovskite structure of La Ferrite (LaFe1-X TiXO3), Journal of Alloys and Compounds, 2020,845, 155040.

[104] Sasikala C, Suresh G, Durairaj N, Baskaran I, Sathyaseelan B, Manikandan E, Srinivasan R, Moodley MK, Chemical, Morphological, Structural, Optical, and Magnetic Properties of Transition Metal Titanium (Ti)-Doped LaFeO3 Nanoparticles, Journal of Superconductivity and Novel Magnetism, 2019,Volume:32 (6),PP: 1791-1797.

[105] C. Sasikala, N. Durairaj, I.Baskaran, B. Sathyaseelan, M.Heninie, E. Manikandan, fTransition metal titanium (Ti) doped LaFeO3 nanoparticles for enhanced optical structural and magnetic properties, Journal of Alloys and Compounds, year 2017,Volume 712,Page No 870-877.

[106] Bhukal, S., Namgyal, T., Mor, S., Bansal, S. and Singhal, S. Structural, Electrical, Optical and Magnetic Properties of Chromium Substituted Co-Zn Nanoferrites Co0.6Zn0.4CrxFe2–xO4 ($0 \leq x \leq 1.0$) Prepared via Sol-Gel Auto-Combustion Method. Journal of Molecular Structure, 2012; 1012:162-167.

[107] Smart, J.S. The Néel Theory of Ferrimagnetism. American Journal of Physics, 1955;23: 356-370.

[108] Raut, A.V., Barkule, R.S., Shengule, D.R. and Jadhav, K.M. Synthesis, Structural Investigation and Magnetic Properties of Zn2+ Substituted Cobalt Ferrite Nanoparticles Prepared by the Sol-Gel Auto-Combustion Technique. Journal of Magnetism and Magnetic Materials, 2014;358: 87-92.

[109] Berkowwitz, A.E. and Kneller, E. Magnetism and Metallurgy. 1654; Vol.1:Academic Press, New York, 295.

[110] Ibusuki, T., Kojima, S., Kitakami, O. and Shimada, Y. Magnetic Anisotropy and Behaviors of Fe Nanoparticles. IEEE Transactions on Magnetics, 2001;37: 2223-2225.

[111] Bhukal, S., Namgyal, T., Mor, S., Bansal, S. and Singhal, S. Structural, Electrical, Optical and Magnetic Properties of Chromium Substituted Co-Zn Nanoferrites Co0.6Zn0.4CrxFe2–xO4 ($0 \leq x \leq 1.0$) Prepared via Sol-Gel Auto-Combustion Method. Journal of Molecular Structure, 2012; 1012:162-167.

[112] Su, M., He, C., Sharma, V. K., Asi, M. A., Xia, D., Li, X.-z., Xiong, Y. (2012). Mesoporous zinc ferrite:

synthesis, characterization, and photocatalytic activity with H2O2/ visible light. Journal of Hazardous materials, 211, 95-103.,

[113] Du, C.-J., Bu, F.-X., Jiang, D.-M., Zhang, Q.-H., & Jiang, J.-S. (2013). Prussian blue analogue K2Zn3 [Fe (CN)6]2 quasi square microplates: large-scale synthesis and their thermal conversion into a magnetic nanoporous ZnFe2– xO4–ZnO composite. CrystEngComm, 15(48), 10597-10603.

[114] Hayashi, H., & Hakuta, Y. (2010). Hydrothermal synthesis of metal oxide nanoparticles in supercritical water. Materials, 3(7), 3794-3817.

[115] Ramaswamy V, Jagtap N.B, Vijayanand S, Bhange S, Awati P. Photocatalytic decomposition of methylene blue on nanocrystalline titania prepared by different methods. Mater. Res. Bull, 2008; 43: 1145-1152.

[116] Syoufian A, Nakashima K. Degradation of methylene blue in aqueous dispersion of hollow titaniaphotocatalyst: optimization of reaction by peroxydisulfate electron scavenger. J. Colloid Interface Sci, 2007; 313: 213-218.

[117] N. Sanpo, C. Wen, C. Berndt, J. Wang, Antibacterial properties of spinel ferrite nanoparticles. Microbial pathogens and strategies for combating them, in: A. Mendez-Vilas (Ed.), Microbial Pathogens and Strategies for Combating Them: Science, Technology and Education, Formatex Research Center, Spain, 2013, pp. 239-250.